Sustainable Development in Civil and Electrical Engineering (SDCEE 2021)

IOP Conference Series: Materials Science and Engineering
Volume 1236

Online
17 - 19 December 2021

ISBN: 978-1-7138-5437-1
ISSN: 1757-8981

Printed from e-media with permission by:

Curran Associates, Inc.
57 Morehouse Lane
Red Hook, NY 12571

Some format issues inherent in the e-media version may also appear in this print version.

Institute of Physics
Dirac House, Temple Back
Bristol BS1 6BE UK

Phone: 44 1 17 929 7481
Fax: 44 1 17 920 0979

techtracking@iop.org

Additional copies of this publication are available from:

Curran Associates, Inc.
57 Morehouse Lane
Red Hook, NY 12571 USA
Phone: 845-758-0400
Fax: 845-758-2633
Email: curran@proceedings.com
Web: www.proceedings.com

TABLE OF CONTENTS

IOP Conf. Series: Materials Science and Engineering 1236 (2022) 011001 doi:10.1088/1757-899X/1236/1/011001

Preface

The need of the era is sustainable development. Sustainable development is the idea that human societies must live, and meet their needs, without compromising the ability of future generations to achieve their requirements. Sustainability is a multidimensional concept that requires a balance of economic, social and environmental equities of development. For engineering processes, this balance can be achieved by ensuring efficiency in resource use and by reducing the environmental impact, without ignoring the technical, technological and financial concerns related to the process. The civil engineering practitioners need to focus on sustainable development issues by emphasizing on the use of green and high-performance materials, applying research and development results timely into practical projects, thus reducing the use of raw materials in manufacturing, as well as treating the waste causing adverse impact on the global environment. In this context, the research studies on sustainability related issues, e.g. (i) application of alternative materials, (ii) material reuse and recycling, (iii) development of environmentally friendly techniques, (iv) efficient use of the space, (v) reuse of structures or its components, could provide good solution. Sustainable development study of civil engineering can guide in actively choosing and taking appropriate measures in construction stage according to the on-site conditions to reduce adverse effects to ecosystem, reducing the inconvenience and trouble to the nearby, and using scientific methods to strengthen the management of construction workers and improve their safety awareness.

Our special thanks go to IOP Publishing (IOP Conference Series: Materials Science and Engineering) as our publication partner to publish this volume of conference proceeding of SDCEE-2021, the 1st International Online Conference on Sustainable Development in the areas of Civil Engineering and Electrical Engineering, entitled *Sustainable Development in Structural & Transportation Engineering* in their prestigious IOP Conference Series. The book covers a variety of multidisciplinary chapters, which would prove to be useful to students, working professionals, practitioners and researchers.

Kurukshetra, India — **Ashwani Jain**
Hamirpur, India — **Rakesh Kumar Dutta**
Kurukshetra, India — **Chalak Hanuman Devidas**

IOP Conf. Series: Materials Science and Engineering 1236 (2022) 011001 doi:10.1088/1757-899X/1236/1/011001

Conference or Sponsor Logo

IOP Conf. Series: Materials Science and Engineering 1236 (2022) 011001 doi:10.1088/1757-899X/1236/1/011001

List of Committees

PATRON

Dr. Satish Kumar
Director, NIT Kurukshetra

CO-PATRON

Dr. Mahesh Pal
Dean(R&C) & Professor Civil Engg. Deptt.

ORGANISING CHAIR

Dr. K K Singh
Professor & Head, Civil Engg. Deptt.

ORGANISING SECRETARIES

Dr. Ashwani Kumar
Professor & Head, Electrical Engg. Deptt.
Dr. Ashwani Jain
Professor, Civil Engg. Deptt.
Dr. H D Chalak
Assistant Professor, Civil Engg. Deptt.
Dr. Atma Ram Gupta
Assistant Professor, Electrical Engg. Deptt.

INTERNAL ADVISORY COMMITTEE

Dr. A Swarup, Professor, EED, NIT Kurukshetra
Dr. K S Sandhu, Professor, EED, NIT Kurukshetra
Dr. Ratna Dahiya, Professor, EED, NIT Kurukshetra
Dr. Baldev Setia, Professor, CED, NIT Kurukshetra
Dr. S K Madan, Professor, CED, NIT Kurukshetra
Dr. S K Patidar, Professor, CED, NIT Kurukshetra
Dr. Praveen Aggarwal, Professor, CED, NIT Kurukshetra

EXTERNAL ADVISORY COMMITTEE

Dr. Mehar Prasad, IIT Madras
Dr. Anupam Chakrabarti, IIT Roorkee
Dr. Abdul Hamid, Adelaide University, Australia
Dr. Mohamed-Ouejdi Belarbi, Biskra, Algeria
Dr. Ashraf M. Zenkour, Kafrelsheikh University, Iran
Dr. Mirzi Betasolo, PNG University, Papua New Guinea
Dr. V P Singh, Texas A&M University, USA
Dr. C S P Ojha, IIT Roorkee
Dr. Mukand Singh Babel, A. I. T., Thailand
Dr. Ian Holman, Cranfield University, UK.
Dr. Adebayo J. Adeloye, Heriot Watt University, UK
Dr. Rajagopalan Balaji, University of Colorado, Boulder
Dr. K C Patra, NIT Rourkela
Dr. S K Singh, NIH, Roorkee
Dr. Suby Bhattacharya, University of Surrey, UK
Dr. Sanjay Kumar Shukla, E. Cowan Uni., Australia
Dr. Goudappa R. Dodagoudar, IIT Madras
Dr. Debasish Roy, IIT Kharagpur
Dr. Anil Dikshit, IIT Bombay
Dr. Ligy Philip, IIT Madras
Dr. Albert Butare, CEO: Africa Energy Services Group
Dr. Faisal Husain, Univ. of Washington, Seattle, USA
Dr. Satish Chandra, IIT Roorkee
Dr. R D Gupta, MNNIT Allahabad
Dr. R K Dutta, NIT Hamirpur

IOP Conf. Series: Materials Science and Engineering 1236 (2022) 011001 doi:10.1088/1757-899X/1236/1/011001

Dr. Rajesh Gupta, , MNNIT Allahabad
Dr. Niranjan Kumar, , NIT Jamshedpur
Dr. P S Bhowmik NIT Durgapur
Dr. Arup Goswami, , NIT Silchar
Dr. Rohit Bhakar, MNIT Jaipur
Dr. Y P Verma, UIET Chandigarh
Dr. O. P Rahi, NIT Hamirpur
Dr. Prerna Jain, NIT Jaipur
Dr. Bhavnesh Kumar, NSIT Delhi
Dr. Baidynath Bag, NIT Raipur
Dr. T Ramesh, NIT Andhra Pradesh
Dr. S Karmakar, NIT Rourkela
Dr. Mrinal Kanti Sarkar, IIT Manipur
Dr. Giribabu Dyanamina, MANIT Bhopal
Dr. M P R Prasad, NIT Kurukshetra
Dr. Jitender Singh, NIT Jamshedpur
Dr. Om Hari Gupta, NIT Jamshedpur
Dr. Bhanu Pratap, NIT Kurukshetra
Dr. Shashi Bhushan Singh, NIT Kurukshetra
Dr. Pradeep Kumar, NIT Kurukshetra
Dr. Shivam, NIT Kurukshetra
Dr. Dhermender Dheer, NIT Patna
Dr. Aeidapu Mahesh, NIT Surat
Dr. Aashish Kumar Bohre, NIT Durgapur
Dr. Nishant Kumar, IIT Jodhpur
Dr. V V S N Murty, Engineers India Limited
Dr. Manish Kumar, Central University of Haryana
Dr. Muralidhar Nayak Bhukya, CUH
Dr. Sanjeev Gupta, GCET, Jammu
Dr. Mirzi Betasolo, Papua New Guinea University of Technology, Lae, Papua New Guinea
Dr. Pankaj Mani, NIH Patna
Dr. S K Singh, NIH Rooree
Dr. Nivedita Singh, ITM University Gwalior
Dr. Ajay K. Prabhakar, NIT Kurukshetra
Dr. Paratibha Aggarwal, NIT Kurukshetra
Dr. V S Dogra, NIT Hamirpur
Dr. Anil Sachan, MNNIT Allahabad
Dr. Babita Saini, NIT Kurukshetra
Dr. Nitish Puri, GCET, Greater Noida
Dr. Praveen Aggarwal, NIT Kurukshetra
Dr. Sandeep Singla, RIMT University, Mandi Gobindgarh
Dr. Jitendra Singh Yadav, NIT Hamirpur
Dr. Arpan Mehar, NIT Warangal
Dr. A. D. Prasad, NIT Raipur
Dr. Abhilash Nair, NIFFT Ranchi
Dr. Pralhad Walvekar, ZCOER Pune
Dr. Aman Garg, NIT Kurukshetra
Dr. Vikas Pratap Singh, NIT, Uttarakhand
Dr. Jitendra Kumar, NIT Jamshedpur
Dr. Aditya Anupam, NIT Uttarakhand
Dr. Atteshamuddin Sayyad, SRES COE Kopragaon
Dr. Senthil Kasilingam, Dr B R Ambedkar NIT Jalandhar

IOP Conf. Series: Materials Science and Engineering 1236 (2022) 011001 doi:10.1088/1757-899X/1236/1/011001

Dr. Tushar Kanti Dey, NITTR Kolkatta
Dr. Kunal Jain, Punjabi university, Patiala
Dr. Maninder Singh, Punjabi university, Patiala

Peer Review Statement

All papers published in this volume have been reviewed through processes administered by the Editors. Reviews were conducted by expert referees to the professional and scientific standards expected of a proceedings journal published by IOP Publishing.

- **Type of peer review:** Double Anonymous
- **Conference submission management system:** Morressier
- **Number of submissions received:** 55
- **Number of submissions sent for review:** 36
- **Number of submissions accepted:** 15
- **Acceptance Rate (Submissions Accepted / Submissions Received × 100):** 27.3
- **Average number of reviews per paper:** 2.27
- **Total number of reviewers involved:** 27
- **Contact person for queries:**
  Name: H D Chalak
  Email: chalakhd@nitkkr.ac.in
  Affiliation: National Institute of Technology Kurukshetra - Civil Engineering

IOP Conf. Series: Materials Science and Engineering 1236 (2022) 012001 doi:10.1088/1757-899X/1236/1/012001

Heat Transfer Enhancement in Composite Building Wall using Phase Change Material

Dhanusiya G[1[0000-1111-2222-3333]], Mohit Gulati[1], Rajasekaran T[2], Elavarasan G[1], Balamurugan A[1]

[1] SRM Institute of Science & Technology, Delhi NCR Campus, Uttar Pradesh, India, 201204
[2] SRM Institute of Science & Technology, KTR Campus, Tamil Nadu, India, 603203
er.bdhanusiya@gmail.com

Abstract. Phase Change Materials (PCMs) are capable of reducing the thermal load transfer rate between outdoors and indoors and maintain stability in the indoor temperature when used in building applications. In this study, a variation of layers of two Bio-PCMs integrated with layers of general construction material and insulation have been analyses discretely as well as in combination. The observations from various models have been compared to draw the result and perceive the best model of a Bio-PCM wall for constructing thermally efficient buildings. The wall models have been created and analyzed through Computer Aided Design (CAD) modeling and simulation techniques using Transient Thermal module in ANSYS software. Coconut fat and Palm fat being highly efficient and easily available Bio-PCMs, are used in this study and attractive results have been obtained from them.

1. Introduction

The world is currently facing a climate crisis. Earth's average temperature is on a steady rise. Human activities have already raised earth's average temperature by 1°C and there is a high certainty that the 2°C mark will also be breached soon. The rising temperatures are abominable for human comfort. Current HVAC systems used for maintaining the indoor environment comfortable for humans have not solved the problem but only added to the severity of the scenario. The Phase Change Materials have the properties of reducing the rate of thermal load transfer. Several studies have been conducted on the Inorganic PCMs using them in making the buildings thermally efficient and reduce fluctuations in indoor temperatures and have obtained mesmerizing results. The most promising results were obtained from Paraffin based PCMs. But, the paraffin based PCM is a petroleum based commodity. Petroleum reserves are already over consumed and are about to extinguish soon. Thus, PCM based technology would be sustainable only if the supplies for the PCM are easily replenish able. This draws the interest of this study towards the highly available and replenishable type of PCM i.e. Bio-PCMs. Bio-PCMs are phase change materials derived from natural sources such as bi-products of trees, animal fat and even organic wastes like vegetable peels and feed stock. Bio-PCMs also have the property of absorbing the heat and using it in changing its state like the various other paraffin based PCMs. Among the Bio-PCMs, Coconut fat and Palm fat attain most of the merits required for a PCM to be used in thermally efficient building applications. They both have a melting point near the human comfort temperatures and have a source which has a high availability and are replenishable. Palm tree and Coconut tree can be grown easily near coastal areas and the use of their products as PCM can generate a new source of income for farmers and will generate employment in the PCM processing plants. Not many studies have been conducted on the utilization of Bio-PCMs in the building application. Some of the studies and their results are stated. [1-4] studied the application of different types of PCMs in the roofs of buildings and focused on the implementation of various design of the roof for better latent storage. The following implementations showcased the reduced heat flux through the roofing complex. [5, 6] tested the PCM equipped wallboards for the use in thermally efficient building applications. These studies provide us with an understanding that how PCM technology can be transformed into an enhanced commercial model in the form of ready to install wallboards. Researchers have incorporated the PCM into hollow bricks and blocks and simulated the latent storage characteristics of the model [7-9]. These types of design proved to be beneficial in lightweight construction along with the efficient performance of the PCM. Advanced models of composite walls with PCM blended into the composition of construction materials, provided the PCM composite with a shape stability, mechanical strength and latent storage characteristics [10, 11]. The simulations on the models concluded that the model sticks well with the latent storage characteristics of the PCM blended in it. [12-14] Researchers have conducted various experiments using the wall models having the PCM as an integral layer. It was observed that the PCM performed most efficiently in this setup. [15, 16] experimentally analyzed scaled down constructed models of a room with walls and ceiling integrated with PCM layers. The analysis was carried out considering the directions of the wall i.e. North, East, West and South along with a time variable for 24 hours of recorded data of the performance of the setup. This type of analysis presents an impressive practical approach. [17] compared the bio-based PCMs as a substitute to paraffin PCMs and stated that Bio PCMs can absorb, store, and release large quantities of latent heat, comparable to conventional paraffin's, and are equally suitable for microencapsulation. [18] Performed a numerical study to investigate the behaviour of the melting process of Coconut fat based PCM under convection heat transfer condition. The investigation concluded that coconut fat has a melting range around the comfortable temperatures and can be used as a PCM.[19] developed macro-encapsulated palm oil based PCM solution for latent heat storage and compared the solution with the present PCM based solution available in the market. The results of the Palm oil based solution was found to be more attractive for thermal storage in building applications. [20] performed a multipurpose investigation on thermal behaviour and life cycle assessment on expired palm oil from food industry as a Phase Change Material for thermally efficient building applications. This study revealed that the expired palm oil can perform efficiently at temperatures near human comfort and can be used as PCM in building applications for latent heat storage. [21] followed an experimental procedure on various types of palm fruit to determine the thermal characteristics of the oil produced from them. A comparison among the thermal properties of various palm oils was produced as a result of this experimental analyse. It has been very helpful to analyse how PCMs fit into building applications based on the above-mentioned properties and optimization techniques, which have been documented by numerous research studies and articles. In response to this perception, we developed a virtual model of PCM fused in working as a wall-layer material. This study compares different b-PCM thicknesses in order to evaluate the thermal performance of composite building walls. Same-sized composite walls ANSYS transient tool is used to analyse different b-PCM for thermal condi-tions. A multi-layered wall's air cooling and temperature distribution has been studied. This is a case in which palm fat & Coconut fat is used as b1-PCM and b2-PCM. As the temperature rises above a desired level, heat is absorbed, and heat is released when the temperature falls below that level. This study also present the heat reduction rates throughout the day and energy savings resulting from the multi-layered composite wall.

2. Description of the model

Four models of wall with (1m x 1m) cross-section area have been designed using geometrical design module in ANSYS WORKBENCH. The models designed are identical to an extent such that the basic design of a conventional wall is common in all the models, while the models distinguish by the combination of Bio-PCM used in them. The combination of layers used in the wall models are displayed in the Fig.1. along with the orientation.

Fig.1. Representative models of composite wall (a) with b1-PCM & b2-PCM, (b) with b1-PCM, (c) with b2-PCM, (d) without b1& b2 PCM

2.1. Mathematical Modeling

The Prototype walls for the PCM integrated room in addition another room type are used as composite panels. The outer layer of the wall is exposed to the air, which is the goal of convective heat transfer and through the air. In this work, it is assumed that the ambient temperature is equal to the temperature of the outer layer of the composite wall, which shows that the convection is not considered due to heat loss. The heat transfer process between the different layers of the composite wall due to heat conduction is determined by the one-dimensional heat transfer equation based on Fourier's law[22].

$$Q = -kA\frac{dT}{dX} \qquad (1)$$

Where Q is the heat flow, k is the thermal conductivity of a single layer of material, A is the cross-sectional area, dT is the temperature difference between the opposite ends of the wall, dx is the wall thickness. The process is unsteady and the ambient air temperature changes with time. The same is true for the temperature of walls and their corresponding layers, so the process takes the equation that controls time as a variable there is no heat source can transfer heat between the walls, inside the wall can be regarded as one-dimensional. The equation gives a one-dimensional form of the diffusion equation in the unstable state and without internal heat release which can be used for the initial value problems (I.V.P) are given by the eq.(2).

$$\frac{d^2T}{dX^2} = \frac{1}{\propto} \times \frac{dT}{dt} \qquad (2)$$

The initial and boundary conditions for eq.(2) asre as follows:

Initial condition, at 12:00 AM, T_{wall} = 36 °C

Boundary conditions, at X=0, $T_{wall} = T_{amb}$

where X=L, T_{wall} represents the temperature of the wall surface, T_{amb} reperents the ambient temperature and L represents the thickness of the wall.

2.2. Standard wall model

This model defines a conventional wall setup of brick and cement-mortar combined with an insulation setup of aluminium and gypsum wallboard. The standard wall model is a reference for comparing the results obtained from analyzing the various other wall models.

2.3. b1 wall model

A layer of Palm fat (stated as b1) of 10 mm thickness enclosed in an aluminium container having thickness 0.7 mm is combined with the standard wall model to form a b1 PCM wall model.

2.4. b2 wall model

The b2 PCM wall model is designed by using Coconut fat (stated as b2) having thickness of 10 mm enclosed in an aluminium container of 0.7 mm thickness along with the conventional wall layers and insulation.

2.5. b1-b2 combined wall model

As suggested by its name, the b1 and b2 PCM wall models are blended to develop a combination of Palm and Coconut fat layers separated by a 0.7 mm thick aluminium sheet and contained in an aluminium envelope having 0.7 mm thickness. The conventional layers of brick (100 mm thick), cement-mortar (15 mm thick) and gypsum wallboard (12 mm thick) still play their part in the design of this model. Both the PCM layers used are 10mm thick each. Materials used in simulations had specific properties, which are listed in Table 1. Material assignment module receives these properties before meshing has been completed as an input to the material assignment module. Since the behaviour of a material in a process is highly dependent on its properties, these properties are crucial for input. Depending on the type of mesh selected, the number of nodes and elements will be distributed differently. Elements and nodes are generated by the solver according to the model's structural requirements. The chosen mesh is a quad type with a maximum edge length of 5 millimetres (mm). Due to the fact that the walls are symmetrical and have no complex features, the mesh used is a structured mesh. To calculate the results efficiently, the structured mesh is preferred over the unstructured one. A coarse, medium and 4 mm element-sized fine mesh have been generated using all three meshing types. For medium and fine meshing, however, the simulation fails. Interior and exterior layers of the wall model are represented by the diagrams' top and bottom layers, respectively. There are minimum and maximum temperatures developed at different layers of the model over a certain time period, and these temperatures are shown as a temperature contour. On the diagram, the maximum temperature is represented by the dark red layers, while the minimum temperature is shown by the dark blue layers.

Table 1 Thermal properties of the material

Material	**Density (kg/m^3)**	**Conductivity (W/m K)**	**Specific heat (J/kg K)**	**Material**	**Density (kg/m^3)**
Aluminium	2689	237.5	951	Aluminium	2689
Brick	1536	0.75	523	Brick	1536
Cement-mortar	1406	0.3505	1050	Cement-mortar	1406
Coconut oil	920	0.166	3750	Coconut oil	920
Palm oil	885	0.1717	1875	Palm oil	885

3. Results & Discussions

PCM were subjected to a convection-based heat transfer process that moved heat from the air to the outermost layer. Through conduction, the heat was exchanged in the inner layers. Air temperature was consistently considered in general recreation as a factor in fluctuating time. The temperature of the air and the temperature of the model's deepest layer were taken into account over a period of hours. This is done using the ANSYS Transient Thermal Module (TTM). This is a heat transfer measure administered by the numerical condition of time-dependent heat transfer. In the temperature contours, the wallboard layer at the interior is the coldest layer, while the brick layer at the exterior is the hottest.

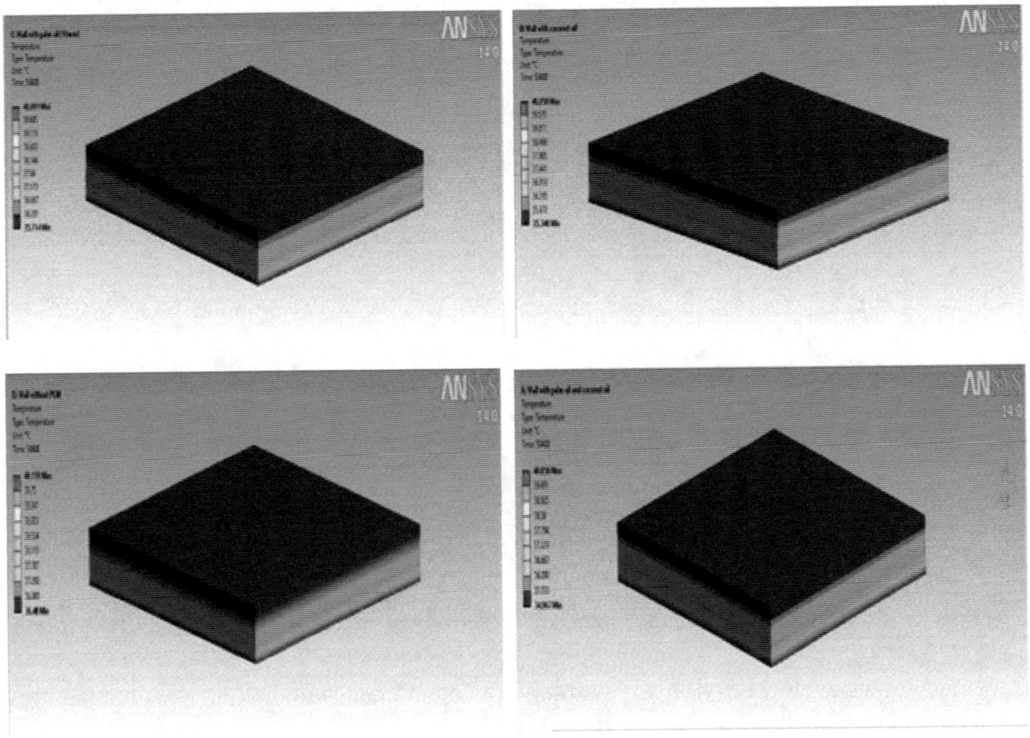

Fig. 2. Thermal contours across the walls

The Fig.2. displays four different temperature contours with a rainbow style colour distribution across the layers of the model walls. The rainbow style colour distribution depicts the temperature variation among the layers. Red colour depicts the hottest layer, then the colour band varies to orange, yellow, green, light blue and finally reduces to dark blue, which depicts the coldest layer among the various layers of the wall model. The model without PCM have contours varying from 40.159°C to 36.48°C, b1 PCM model temperature contour varies from 40.091°C to 35.714°C, b2 PCM model contours varies from 40.058°C to 35.348°C and the temperature contours of b1&b2 combined wall model has achieved a variation from 40.056°C to 34.967°C. This shows the increasing performance in reducing temperature of the topmost layer shown in the fig.2. as the setup is optimized from no PCM to b1 PCM to b2 PCM to both b1&b2 PCM combined. An increased overlap of dark blue contour over light blue contour can also be observed in the contours of the models having PCM. This interprets the increased heat reduction due to PCM when compared with model without PCM.

3.1 The effect of b1 PCM

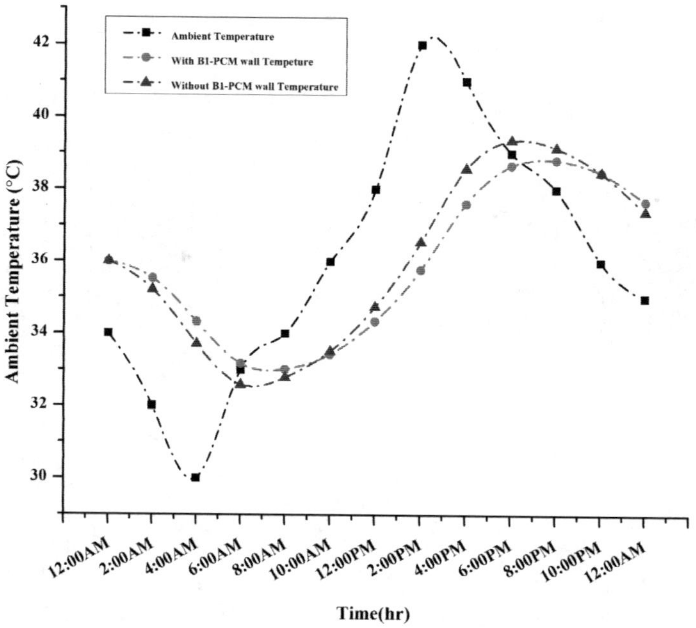

Fig.3. Inner temperature of the composite building wall with b1-PCM and without b1-PCM

The graph in the **Fig.3**. displays the simulation results of the b1 PCM wall model. The line joining the squared points describes the behavior of environmental conditions stated as ambient temperature. The ambient temperature stands at 34°C initially at 12:00 AM. There is a continuous fall in ambient temperature till 4:00 AM after which the ambient temperature starts to rise. The time step of 4:00 AM can be stated as the sunrise. Further the temperature rises till it reaches its peak of 42°C at 2:00 AM that is when the sun is at the top, after which the temperature falls continuously till the end of the observation time period at 12:00 AM. The behavior of the wall model without B1 PCM is comparative to the ambient temperature characteristics as it falls and rises along with the ambient temperature. The temperature differences between them are due to the higher initial temperature of the wall models and the insulation due to the presence of aluminum and gypsum wallboard layers. The behavior of the wall model with b1 PCM layer, depicted in form of line joining circular points shows a similar behavior as the wall without PCM but, it provides an extra smoothening of the curve which is the effect produced solely by the b1 PCM layer. When the temperature of the wall without PCM drops from 36°C to 32.59°C in the time steps ranging from 12:00 AM to 6:00 AM, the wall with B1 PCM layer drops with a smaller slope from 36°C to 33.172°C along the same time steps. After 6:00 AM, the temperature of the wall without PCM starts rising till its peak of 39.34°C and the PCM wall shows a smaller rise in slope than the wall without PCM and thus depicts a lower fluctuation in temperatures.

3.2 The effect of b2 PCM

Fig.4. shows that the graphical result of the simulation performed on wall model with b2 PCM. The behavior of the ambient temperature is presented by the line joining squared points, the behavior of the wall without PCM is represented by line joining triangulated points and the line joining the circular points depicts the behavior of the wall with B2 PCM.

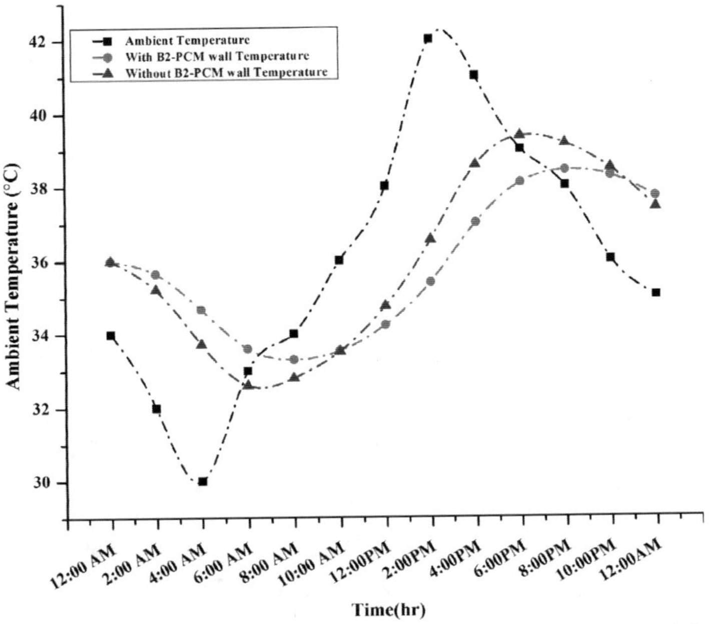

Fig.4. Inner temperature of the composite building wall with b2-PCM and without b2-PCM

The process under observation begins at 12:00 AM at which the ambient temperature is 34°C and the initial temperatures of the wall models being 36°C. In the beginning, the temperature of wall without PCM falls to 33.714°C from 36°Cas there is a fall in ambient temperature in the time steps ranging from 12:00 AM to 4:00 AM. After 4:00 AM, there is a sharp rise in ambient temperature but, the temperature of the wall without PCM has not fallen as much. Thus, an intersection is observed at 6:00 AM. Further, the temperature of the wall without PCM rises along with the ambient temperature and drops along it. On the other hand, the temperature of the wall with PCM wall falls along with the temperature of the wall without PCM but maintains a significant difference in temperature. By the time step of 6:00 AM, the temperature of the wall without PCM drops to 32.59°C, the temperature of the wall with B2 PCM has only dropped to 33.595°C. This point depicts the behavior of the PCM. While the temperature of wall falls below the ambient temperature, the temperature of the wall with B2 PCM stays above the ambient temperature at this time step. This shows that the PCM slows down the heat transfer rate at sufficiently differentiable rates. The same phenomenon can be observed at the time step of 6:00 PM, the temperature of the wall without PCM has reached 39.359°C, the temperature of the wall with PCM is at 38.093°C only. A reduction in fluctuation of temperature is observed in this simulation.

3.3 The effect of b1 & b2 PCM

The Fig.5. is a graphical representation of the results obtained by simulation on the model having both b1 and b2 PCM layers in combination. The ambient temperature curve and the temperature curve of wall without PCM is same as described in the B1 PCM and b2 PCM simulation graphs. The combination of the b1 and b2 PCM layers has further enhanced the fluctuation reduction which can be observed at the bottom peak of the temperature of wall without PCM, where it reached 32.59° C but the temperature of the wall with b1 and b2 PCM has dropped only to 34.093°C from 36°C. This creates a 1.503°C temperature difference between them. Similarly at the upper peak of temperature curve of wall without PCM at 39.359°C, the temperature of the wall with b1 and b2 PCM remains cooler at 37.444°C creating a 1.915°C temperature difference.

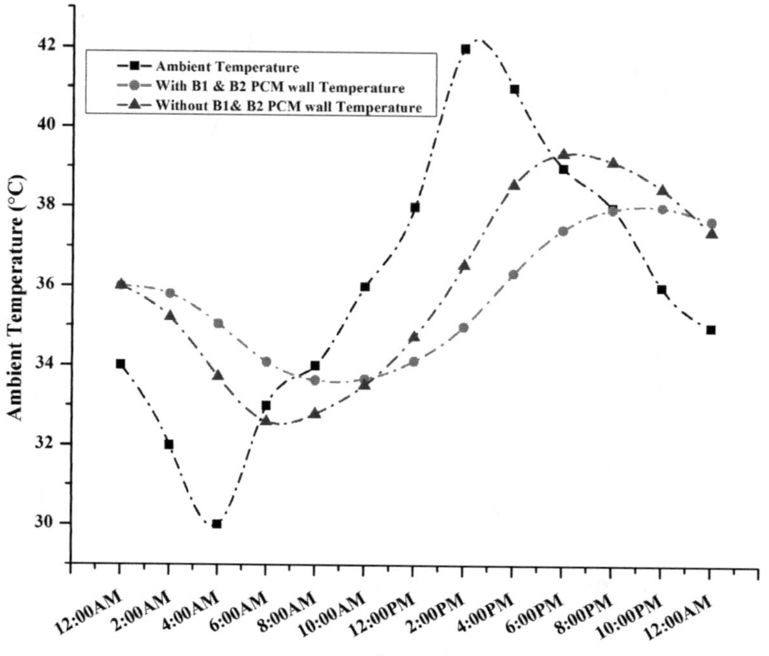

Fig.5. Inner temperature of the composite building wall with b1,b2-PCM and without b1,b2-PCM

4. Conclusion

In this study Different b-PCM has been used in a composite wall to analyse the thermal insulation properties of the material. The PCM wall maintains its thermal inertia for longer periods of time, even as the ambient air temperature decreases. As a result, walls with PCM were found to be more thermally stable than walls without PCM. Due to the addition of PCM, temperature variations are reduced in intensity and their amplitudes are reduced significantly. PCM is most effective from 2 PM to 10 PM. The intersection of the temperature curve of wall without PCM and wall with b1 PCM is observed slightly before 10:00 AM, the intersection of the temperature curve of wall without PCM and wall with b2 PCM is observed exactly at 10:00 AM and the intersection of the temperature curve of wall without PCM and wall with B1 and B2 PCM combined is observed slightly after 10:00 AM. This shows that the b2 PCM produces a higher time delay in temperature change than b1 PCM and the combination of b1 and b2 PCM produces even higher delay than b2 PCM alone. On comparison with the wall without PCM, the difference at the bottom and top peaks of b1 PCM was 0.582°C and 0.696°C respectively, of b2 PCM was 1.005°C and 1.266°C respectively and of b1-b2 PCM combined was 1.503°C and 1.915°C respectively. This clearly derives towards the conclusion that b2 PCM i.e. Coconut fat is a better PCM than b1 PCM i.e. Palm fat and the combination of the two performs the best among them and reduces considerably high amount of temperature fluctuation loads.

References

1. Srinivasan PSS., Ravikumar M. 2014, Heat transfer analysis in PCM filled RCC roof for thermal management, *Journal of Mechanical Science and Technology*, 28(3), 1073-1078.
2. Liu C, Zhou Y, Li D, Meng F, Zheng Y, Liu X. 2016, Numerical analysis of thermal performance of a PCM filled double glazing roof, *Energy and Buildings*, 125, 267-275.

3. Hanchi N, Hamza H, Lahjomri J, Oubarra A. 2017, Thermal behaviour in dynamic regime of a multilayer roof provided with two phase change materials in the case of a local conditioned, *Energy Procedia,* 139, 92-97.
4. Chung M.H, Park J.C. 2016, Development of PCM cool roof system to control urban heat island considering temperature climatic conditions, *Energy and Buildings,* 116, 341-348.
5. Berardi U, Soudian S. 2018, Benefits of Latent thermal energy storage in the retrofit of Canadian high-rise residential buildings, *Build Simul*, 11, 709-723.
6. Soares N, Reinhart C.F, Hajiah A. 2017, Simulation-based analysis of the use of PCM-wallboards to reduce cooling energy demand and peak-loads in low-rise residential heavyweight buildings in Kuwait. *Build Simul*, 10, 481-495.
7. Dabiri S, Mehrpooya M, Nezhad E.G. 2018, Latent and sensible heat analysis of PCM incorporated in a brick for cold and hot climatic conditions, utilizing computational fluid dynamics, *Energy*, 159, 160-171.
8. Aketouane Z, Malha M, Bruneau D, Bah A, Michel B, Asbik M, Ansari O. 2018, Energy saving potential by integrating Phase Change Material into hollow bricks: the case of Moroccan buildings, Tsinghua University Press, *Springer-Verlag GmbH Germany, Springer Nature*
9. Muthuvel S, Saravanasankar S, Sudhakarapandian R. 2014, Numerical and experimental analysis on inorganic Phase Change Material usage in construction, *Springer, The Institution of Engineers* (India) 95(4) 231-238. 2015, 103, 374-386.
10. Kong X, Jie P, Yao C, Liu Y. 2017, Experimental study on thermal performance of phase change material passive and active combined using for building application in winter. *Applied Energy*, 206, 293-302.
11. Fateh A, Klinker F, Brutting M, Weinluder H, Devia F. 2017, Numerical and experimental investigation on an insulation layer with phase change materials (PCMs). *Energy and Buildings,* 153, 231-240.
12. Venkatesh L, Amruth J, Sammeta S.R. 2014, Phase change materials in Building construction to reduce room temperature fluctuations. *International Journal of Mechanical and Production Engineering*, volume -2, issue-7.
13. Li Y, Liang W, Zhou J, Long E. 2017, Experimental study on thermal performance improvement of building envelopes integrated with Phase Change Materials in an air-conditioned room, *Procedia Engineering*, 205, 190-197.
14. Hasan M.I, Basher H.O, Shdhan A.O. 2018, Experimental investigation of phase change materials for insulation of residential buildings, *Sustainable Cities and Society*, 36, 42-58.
15. Derradji L, Hamid A, Zeghmati B, Mohamed A, Bouttout A, Errebai F.B. 2015, Experimental study on the use of Microencapsulated Phase Change Material in walls and roofs for energy savings, *Energy Engineering*, 141(4), 04014046(1-10).
16. Naresh R, Parameshwaran R, Ram V.V. Bio-based Phase change materials, Bio-based Materials and Biotechnologies for Eco-efficient Construction, *Elsevier.*
17. Al-Jethelah M.S.M, Al-Sammarraie A, Tasnim S.H, Mahmud S,Dutta 2018, A. Effect of convection heat transfer on thermal energy storage unit, De Gruyter, 16, 861-867.
18. Landi F, Fabiani C, Pisello A.L. 2020, Palm oil for seasonal thermal energy storage applications in buildings: Thepotential of multiple melting ranges in blends of bio-based fatty acids, *Journal of Energy Storage*, 29, 101431.
19. Fabiani C, Pisello A.L, Barbanera M, Cabeza L.F. , 2020, Palm oil-based bio-PCM for energy efficient building applications:Multipurpose thermal investigation and life cycle assessment, *Journal of Energy Storage* 28, 101129.
20. Alakali J.S, Eze S.O, Ngadi M.O. 2012,Influence of Variety and Processing Methods on SpecificHeat Capacity of Crude Palm Oil, *International Journal of Chemical Engineering and Applications*, Vol. 3, No. 5.

IOP Conf. Series: Materials Science and Engineering 1236 (2022) 012002 doi:10.1088/1757-899X/1236/1/012002

Study of self-compacting concrete containing coal bottom ash and fly ash

Mohd.Adnan[1], Ankit Kumar[1], Kunal Kant[1], Paratibha Aggarwal[1],Yogesh Aggarwal[1]

Civil Engineering Department, National Institute of Technology, Kurukshetra.

Abstract. This paper consists of a detailed analysis of the use of industrial by-products (coal bottom ash and fly ash) as an admixture in the self-compacting concrete. The judicious use of industrial by-products and the proper waste management can lead to a sustainable development in the field of infrastructure. This paper talks about the physical and chemical properties of these by products used in the self-compacting concrete. Critical and definite analysis has been carried out on the reviewed literature and further scope of work in the analysis and interpretation has been addressed. On a positive note, it can be stated that the use of industrial by-products in self-compacting concrete can lead to a sustainable development.

1. Introduction

Concrete is a mixture that is used in the construction of the structural element comprising of various constituents like coarse aggregate, fine aggregate (sand), binding material (cement), water and other admixtures which gives specific qualities for specific purposes. The history of concrete is as old as of the Roman civilization and the structures made in those times are still well to do. The concrete is made so to bind itself with the ribs in the reinforcements and cement plays a vital role due to the hydration phenomenon which takes place sticking together all the rest of the components.

The concrete taken into consideration for the review study is self- compacting concrete with specific admixture. Research and development in the field of high performing concrete has led to the production of self-compacting concrete. The conventional concrete showed problems like uneven compactness which in turn exhibited negative impact on the permeability of concrete. Self-compacting concrete as the name suggests is a variety of concrete which settles down in the structural elements on its own due to its self-weight and also gives high strength and is highly workable. The history of self-compacting concrete can be dated back to 1988 where in Japan it was first developed. The outstanding deformability and segregation resistance properties have led to the increase in the demand of Self Compacting Concrete. SCC is very identical to the Normally Vibrated Concrete (NVC) with just an increase in the amount of additives like Coal Bottom Ash (CBA) and Coal Fly Ash (CFA) as a replacement to fine aggregates and Portland cement respectively.

CBA is a by-product in the combustion of coal which has not been used judiciously in the construction sector. CBA can replace the purpose of fine aggregates (namely sand) in concrete due to qualities like hardness and light weight.CBA majorly consists of silica, alumina and iron with mild traces of magnesium, calcium and other chemical sulphates. It can be compared to the sand present in the river beds in terms of size. Fresh concretes formed using CBA are quite active against water loss due to bleeding and also cause low deformation due to plastic shrinkage.On the other hand CFA is an alternative to the binding materials (namely cement) used in the construction sector which impart characteristics like increase in long term strength, decrease in drying shrinkage of surface area and suppression of alkali-silica reactions. CFA is also obtained as a by-product in the coal combustion reaction similar to CBA. Disposal of CFA is generally not done in a proper way creating hazard for the environment and also raising the chances of groundwater pollution which may lead to other serious health issues in the society. Thus we can say that concrete production using CFA might be an alternate method to reduce the chances of environmental and social issues. Although CFA is being used as a constituent in the manufacture of Pozzolanic Cement, CBA is still not used at such a high rate due to its less availability.

Environmental safety has always been a concern for a healthy planet and to establish sustainable development. The rapid pace of industrialization and urbanization has led to increase in the global warming due to the improper disposal of the by-products produced and lack of waste management. There has been a considerable increase in the level of by-products in the recent years. It is high time sustainable development has become the need of the hour to maintain the ecological balance in the nature. CFA produced from the factories have a severe impact on the environment when fumed in the atmosphere as it settles down on the leaves of the plants and interferes with the life processes like photosynthesis leading to global warming. They also interfere with the health of human beings as the emission of coal ashes namely CBA and CFA can lead to skin, lungs and bladder cancer and are grouped under the category of Group-I human carcinogen. Developed countries having advanced technologies like Malaysia have been using the residual CBA and CFA from their thermal powerplant as a major proportion in their concrete production. Their thermal power plants have been producing around 20-80% of CBA and CFA which is a huge amount in the case for manufacturing concrete for infrastructure use. In 2015 the use of fly ash in the production purposes was around 43% in India, whereas, it was as high as 70 % in China and around 53% in USA.

The advantages of Self Compacting Concrete (SCC) over Normally Vibrated Concrete (NVC) is the elimination of external or internal vibration for compaction, comparatively better workability and flow ability as well as better bonding with the reinforcement. It provides high durability, strength and smoother aesthetic finishes which leads to its use in architectural and repairing works. Also use of SCC leads to improved construction ability and minimizes the voids present in the highly reinforced areas. This inturn reduces the permeability of the concrete. The drawback observed in the use of SCC is the high cost as the admixtures or by-products constituents used in its production are quite expensive for construction purposes. Also a large number of trial batches and laboratory tests need to be conducted to bring in use a well-designed mix. Another disadvantage observed is that we do not have a specific designed code for the use of SCC as we have for NVC.

The major aim for this study is to compare the self-compacting concrete with conventionally used concrete on these basis i.e. fresh propertiesand mechanical properties including various criteria such as slump value, compressive strength, tensile strength, and many other minute criteria taken into consideration for the construction purpose. Both the CBA and CFA as a replacement to sand and cement respectively to clearly judge the changes in the properties have been taken separately.

This paper brings in the comparative study of the various strengths of the papers published by other authors. It will give the idea of utilising self compacting concrete with the mentioned ingredients for the construction of infrastructure and would promote the sustainability in the resources of our nation. This would bring in the practicality in the use of self compacting concrete on the basis of the location it is being used at.

2. MATERIAL AND MIX PROPORTION

The main components of the mix was cement, course aggregate, water, fine aggregates with replacement of the fine aggregates with partial amount of CBA as observed in some of the review papers and also replacement of cement with calculated percentage of CFA in certain other review literature. The test samples in every literature examined consisted of the base sample with no by-products added to it followed by an increase in its percentage.

2.1. CBA

2.1.1. Physical Properties:

After observations in the laboratory is has been noted that CBA is a dark grey colored material mostly angular in shape. It has a major property of light weight and is also porous in nature with a rough texture. The size of the grains plays a major role in determining the characteristics of CBA. It has been observed by studying the CBA produced by various thermal power plants. The grain size analysis of CBA is very similar to that of fi e aggregates with a low percentage of silty clay. Other physical parameters like water absorption, specific density, fineness modulus, bulk density and others have been given in the table below.

2.1.2. Chemical Properties:

There has been a critical analysis done in previous papers using X-ray energy dispersive spectrometry (EDS). It was observed that CBA is mainly composed of silica and alumina. The chemical composition represents that it resembles to the class F materials meaning that more than 70% of CBA is formed from raw or calcinated pozzolanic materials. Other instances are that the loss of ignition is around 8% which also shows that CBA belongs to class F materials group. Traces of unburnt carbons improve the concrete. Overall it can be stated that CBA lies more or less in between fine aggregates and class F materials. Under the chemical analysis performed on CBA traces of copper, zinc, arsenic, nickel, mercury and other harmful elements have also been found.

2.2. CFA

2.2.1. Physical properties:

Unlike CBA, CFA is spherical and round in shape may it be hollow or solid. It is in fine powdered form and orange to deep red in nature due to presence of unburnt carbon and iron. The specific gravity of CFA varies in a wide range from 1.6 to 31, but majorly found around 2.0. this vast variation is due to various factors like shape, uniformity of particles and also due to the chemical composition. On a contrary, it has a low bulk density and high specific surface area to its fine particles. On performing the sieve analysis it was observed that its size can be classified as sandy silt to Silty sand.

2.2.2. Chemical properties:

CFA has a vast number of minerals present in it making it one of the most complex compounds. It is mainly formed of oxides of silica, alumina and calcium, thus having a wide range of pH values ranging from acidic to alkaline. CFA comes under two chemical types on the basis of their industrial uses namely class C and Class F. Loss of ignition of CFA varies from less than or equal to 6% and up to 12% based on the performance and use of the fly ash.

3. TEST METHODS

Various kinds of engineering tests are conducted on the sample like slump test to measure the workability, density test, thermal conductivity test, the degree of water absorption and the test for permeability and porosity is also conducted on the sample. The strength test or the compressibility test taken into account is done for determining the cube strength on a concrete block of 150mm x 150mm x 150mm at 3, 7, 28, 90, 180 days, with each test conducted for three samples and then the mean strength taken as the final compressive strength.

Some of the major tests being conducted on the specimen concrete are:

3.1. Slump Cone Test:
This test is conducted using a frustum shaped iron frame in which the concrete mix is poured and its consistency or fluidity is measured. It has a top diameter of 10cm and bottom diameter of 20cm with the concrete fill up to a height of 30cm when filled. Also a tampering rod is used to settle the mix in the cone. The fall in the concrete mix when the cone is removed is called the slump which then categorizes the concrete having very low, low, medium or high workability. The results give the characteristic of the concrete in addition to the slump value. The various kinds of slump that can be seen are true slump, shear slump or collapse slump.

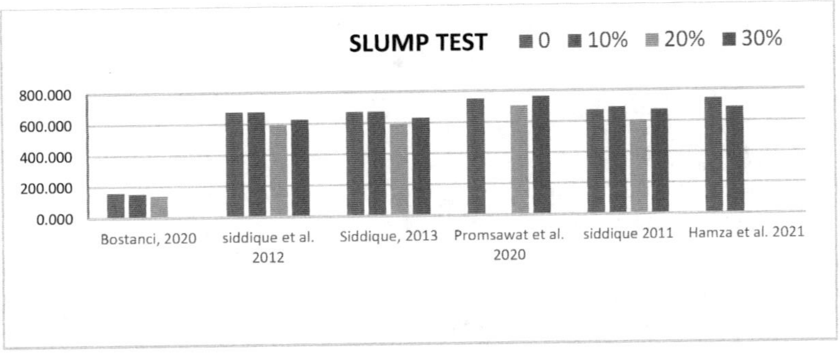

Fig.1. Slump values

3.2. L box Test:
This Japanese test designed by Peterson especially for self-compacting concrete measures the extent of the flow of concrete through the reinforcements. The apparatus is an L shaped rectangular box section with a gate dividing both the sections. The concrete is poured in the vertical column having reinforcements of diameter 20mm and the gate is lifted to let the concrete flow to the horizontal section. The height of concrete is measured at a distance of 200mmand 400mm in the horizontal section and the respective timings are also noted. The height of concrete in the horizontal section is represented as proportion of the left concrete in the vertical section which indicates the slope of concrete at rest condition.

3.3. J ring Test
It is a test specifically designed for self-compacting concrete to check the ability of the concrete to pass through. It consists of a base plate and a ring with vertical rod like projections. The ring has a diameter of 300mm with projections at an interval of 48mm. this test requires the use of the slump cone which is filled with concrete mix of around 6 litres. But in this case the tamping is not done instead we just strike off the concrete level using a trowel. The cone is lifted up and the concrete is left to flow through the j ring. The average diameter and the height difference in just inside and outside of the J ring are calculated. The acceptable height is between 0 and 10 mm.

3.4. V funnel Test
It is a Japanese test for self-compacting concrete used to find the flow ability of the concrete consisting of 20mm aggregates. The funnel is of 490mm at the top entry and of 70mm at the neck or end of the funnel. The trap door of the apparatus is kept close and around 12 litres of concrete is filled in it and after a period of 10 seconds the door is opened and concrete is left to flow under gravity. The total flow time is noted down.

3.5. U box Test

It is also a SCC based test which is used to measure its filling ability. It consists of u shaped box with a wall dividing it into two compartments. The first compartment is filled with around 20 litres of concrete and is left to stand for 1 minute. The movable gate is then lifted up and the concrete is left to flow to the other compartment until it settles. The height of concrete column is measured in both the compartments and the difference is noted down. It shows the flowability and passing ability of the concrete.

3.6. Compression Test

This is the most commonly conducted test in the laboratory consisting of a cube of dimension 150mm x 150mm x 150mm filled with concrete mix andleft to cure for specific number of days. The days of curing taken into account for the self-compacting concrete are 3, 7, 28, 90 and 180 days. After completion of curing the specimens are dried and then taken for compressive test in Universal Testing Machine (UTM).

3.7. Splitting Tensile Strength Test

This test is an important part of the concrete testing as structural loads make concrete critical in tensile failure. The tensile strength of concrete is less than the compressive strength around 10%.Cylindrical samples of concrete of diameter 15 cm and height 30 cm is casted and placed under the compression testing machine with plywood sheets at top and bottom of the specimen. A total load of 9.9 ton/minute to 14.85 ton/minute is applied on the specimen and the breaking load in noted.

4. RESULTS:

4.1. Fresh Properties:

Workability is one of the most basic and just criterion for measuring the quality of concrete mix. Tests specifically designed for SCC are slump flow test, L-Box test, V funnel test, J ring test and U box test. All the tests conducted have been majorly done on 5%, 10%, 20% and 30% of CFA or CBA replacing the cement or aggregates. In an extensive report by Siddique (2012) all the three major tests i.e. slump cone test, L box test and V funnel test were conducted. The results show a specific trend that with the increase in the portion of CBA and keeping the percentage of fly ash varying for each data set from 4% to around 10% there was a decrease in the slump value from 675mm to 625mm. The internal trend with varying fly ash content was also the same and a decline of around 10% was noticed. Esquinas et al. (2018) studied mechanical behavior of fly ash and observed that with a combination of super plasticizers with fly ash there was a decrease in the slump value. It also included the test results of L box test, V funnel test and J ring test. The results of the V funnel test consistently increased which meant that with increase in the amount of fly ash combined with super plasticizer the workability also increases. Similar results were found for the J ring test as well where the radius of flow circle kept on reducing. Siddique (2013) gave the result that with keeping the percentage of fly and super plasticizer constant and varying the amount of CBA present there was a decrease in the slump value at a point when the percentage of CBA was 0.078% where the slump value was noticed to be 591 but when the value of CBA was further increases to 0.116% an unusual phenomenon was noticed where the slump value increased again to 627. L box test, V funnel test and J ring test was also performed in regard with this paper. And the results were similar to that obtained in the slump flow test in which the best sample was found at a point where CBA percentage was 0.078%. But in case of J ring test it was seen that the results kept on increasing with increase in the amount of CBA in the mix. At 0.078% the height difference was noted to be 4.7 mm which is well within the limits but as we moved to CBA percentage as 0.116% the difference came out to be 11.6 mm which is not at all acceptable in case of SCC. Khongpermgoson et al. (2020) studied that slump for the base sample without any CBA proportion was 180mm and steeply decreased to 0mm in a matter of around 90 minutes. Furthermore fore samples of SCC were taken with the amount of CBA as 15%, 25%, 35% and 45% and the respective values of slump were noted. It was observed that the slump value was 700mm in case of 15% replacement which reduced to 680mm for 25% but again increased to 700mm for the rest of the samples. So the findings can be concluded with concrete mix with 25% replacement with CBA as the most optimum and efficient concrete mix. Promsawat et al.(2020) experimented with a combination of fine aggregates and cement. Three proportions of fine aggregates were taken with each proportion containing three proportions of cement as 5%, 10% and 20%. Slump cone test and V funnel test were conducted for all these samples. The slump value for the base mix was 750mm. A particular pattern was noted as with increase in the amount of fine aggregates slump value increased but with the increase in the proportion of cement from 5% to 20% the value of slump decreased. The sample with 20% fine aggregates and 20% cement could be considered as the most optimum and efficient sample as the slump value was noted as 653mm which is minimum of all the observed values. The V funnel test was very similar to that of the slump cone test. The base value was 11.59 seconds. The value of the most efficient sample could be taken as 30% fine aggregates and 5% cement as its value was noted as 9.51seconds and the desirable value for a SCC is around 10 seconds.

Mustapha et al. (2021) studied the slump cone test was performed for the mixture. The base mix gave a result of 550mm slump value and on keeping the percentage of silica fumes constant at 10% and steadily increasing the percentage of fly ash from 25% to 65% there is an increase in the slump value from 600mm to 650mm. L box test and V funnel test were also conducted for the test samples in this paper. The V funnel test results claimed that there is an increase in the time of flow for the selected test samples from a value of 12 seconds for the base mix to 10.2 seconds for the mix containing 65% flyash. Hamza et al. (2021) studied use of Bagasse ash in addition to rice husk. The results were such that the base mix having no amount of admixture had a slump value of 740mm and with a constant increase of ash percentage with an increase of 5% the slump value kept on reducing and reduced to 660mm at 15% ash percentage. In addition to this L box test and v funnel test was also carried out. The results were similar with respect to L box test i.e. a declining curve of the ratio from 0.99 to 0.82. The most suitable mix as in the case of V funnel test was found out to be at 10% replacement which was around 10.19 seconds.

4.2. Mechanical Properties:

The properties which come under the mechanical properties are compressibility test, splitting tensile test and flexural test. Many literatures have been published having the results obtained by performing the tests on SCC. This paper summarizes the trends of all these tests performed on SCC. A study by Siddique (2012) considered 5 samples each of 0%, 10%, 20% and 30%. The 28 days compressive strength for the base mix containing 0% CBA and 4.1% fly ash was found to be 35.25MPa. This value increased when the CBA percentage was increased to 10% to 35.33MPa and then consistently decreased to 25.88MPa for 30% CBA and 4% fly ash. A trend was noticed that with increase in the fly ash percentage the compressive strength kept on decreasing. The tensile strength gave a value of 2.41MPa in case of 0%CBA and 4.1% fly ash which reduced to 2.09MPa at 30% CBA and same percentage of fly ash. The trend noticed was similar to that of compressive strength. Mohammed (2011) stated that while decreasing the amount of cement in the mix and simultaneously increasing the amount of fly ash there is a consistent decrease in the compressive strength of the mix having a change from 22.26MPa at 10% replacement to 8.99MPa at 30% replacement. Also it was seen that at 40% replacement the strength increased to 19.38MPa and then further increased to 22.3MPa for 50% replacement. Sukumar et al. (2006) considered two sets of samples namely series A and B. A series included the mix using only fly ash as an admixture and B series contained fly ash along with quarry dust particles as inert filler. Comparison between the two kinds of sample was done and it is observed that B series showed a comparatively lesser slump value of when compared to A series. Similar trend was observed in the L box test as well where the ratio of the heights reduced in B series when compared to A series.

Siddique (2013) suggested that the maximum strength is actually at the base mixture and when the proportion of CBA is increased and the fly ash proportion is kept constant the strength kept on decreasing. The minimum point up to which the strength decreased was around 25.8 MPa at 30% CBA for 28 days and around 20.8 MPa for 7 days which was around 35.2MPa with 0% CBA at 28 days. Similar kind of results were found with respect to the tensile strength where the base mix strength was 2.4 MPa at 28 days and then consistently decreased to 1.9 MPa. A comparatively higher value of tensile strength was found at 90 days i.e. 2.05MPa which is an increase of around 9%. Singh and Siddique (2014a) studied both compressive strength and splitting tensile strength were taken to consideration. The results stated that with the increase in the proportion of CBA in the sample there is a decrease in the compressive strength. The value of compressive strength was noted to be 38.08MPaof the base mix. The maximum compressive strength was observed at 40% replacement at 28 days where the value obtained was 37.7MPa but was still less than that which could be attained without CBA. The tensile strength showed a completely different pattern where the base mix had a tensile strength of 6.99MPa which showed a curvy pattern in the values with increase in the CBA percentage of the mix. The maximum value was attained at a CBA content of 50% which was 7.96MPa. There was a decrease in the tensile strength at the same level if compared with the 7 day tensile strength which was 8.82MPa. In totality if compared the maximum tensile strength was attained at 100% replacement on 7 day test. Singh and Siddique (2016) studied compressive strength as well as splitting tensile strength. It considered two types of mixes namely A in which the amount of cement was in more proportion and B in which the amount of cement is comparatively less. The compressive strength at 28 days and 90 days for a sample ranging from 20% to 100% replacement with CBA was discussed. The results for sample type A have been discussed at first and then for sample type B. The average 28 day strength was around 38.21 MPa and the average 90 day strength is around 40.83 MPa for first kind of sample and 34.04 MPa and 42.32MPa respectively for the second kind of sample. Taking 28 day tensile strength into consideration the strength comes out to be around 2.67 MPa and around 3.09 MPa at 90 days for sample A and 2.57 MPa and 3.02 MPa respectively for sample B. Bostanci (2020) showed results claiming that the 20% CBA has a compressive strength of 38MPa which decreased from a value of 40.5MPa at 0% CBA for 28 days. Tensile strength test was not included in this paper. Ting (2020) studied 0%, 30%, 40% and 50% CBA. The base mix taken into consideration had a compressive strength of 38.88MPa which decreased to 32.17MPa at 30% CBA value. The 28 days splitting tensile strength was 2.82MPa for 0% CBA which kept on decreasing up to 1.62MPa at 50% CBA. The trend was same for 7 days and 90 days. Promsawat et al. (2020) considered a wide variety of concrete mixes with the base mix having a 28 day compressive strength of 56.25 MPa. The common trend noted for all the mix sets namely 20%, 30% and 50% fine aggregates with cement variation from 5% to 20% i.e. that with increase in cement content the strength decreases. The maximum strength was computed as 61.5 MPa for 20% FA and 5% Cement and tending to a minimum value of 40 MPa for 50% FA and 20% Cement.

Hamza (2020) gave a comparatively lower value of compressive strength i.e. around 26.64MPa in the case of base mix but the trend was similar to the rest of the papers. The values for 15% ash content was 15.22MPa which means that the compressive strength of the mix reduced to around 60% of the base mix on addition of the ash. The tensile strength test was also conducted for the mix and the base mix strength came out to be 2.7MPa at 28 days which firstly increased to 2.71MPa for 5% replacement but then decreased to 1.69MPa at 15% replacement which is around 65% of the base mix.

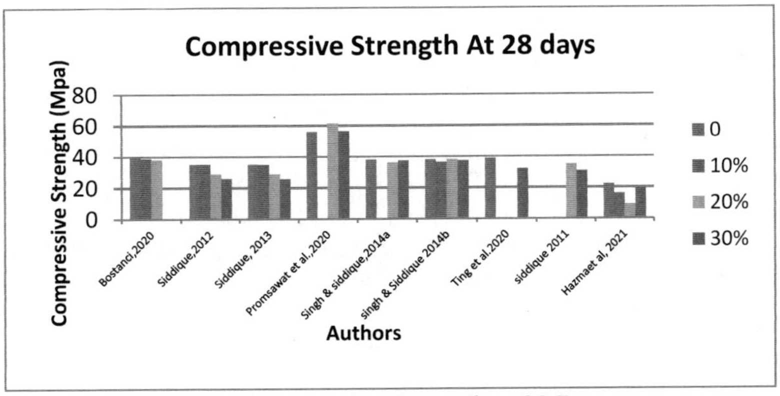

Fig.2. Compressive Strength at 28 Days

The bar chart shown in figure above gives a visual interpretation and differentiation between the compressive strength of wide range of concrete mixes obtained in the various literature taken into consideration. It can be noticed that the results given by Promsawat et al. (2020 has a comparatively higher values of compressive strength but has not included the 10% CBA content. As discussed above the trend that can be seen in the graph is also decreasing thus concluding that with CBA percentage increase the compressive strength of the mix decreases.

The above figure shows the splitting tensile strength comparison of all the papers taken into consideration. As discussed above the tensile strength also follows the same trend as in the case of compressive strength i.e. decreasing slope. The paper having a comparatively unusual trend was by Singh and Siddique (2014b) wherein the graph shows a upward slope on increasing the percentage of CBA in the concrete mix.

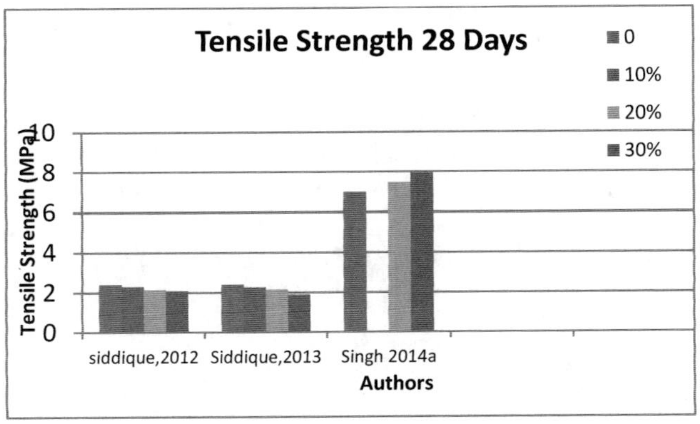

Fig.3. Splitting Tensile Strength at 28 Days

5. Conclusion:

Environmental safety has always been a concern for a healthy planet and to establish sustainable development. The rapid pace of industrialization and urbanization has led to increase in the global warming due to the improper disposal of the by-products produced and lack of waste management. This portion of the paper gives the critical analysis of the various points regarding the use of Coal bottom ash and Fly ash in Self compacting concrete as a replacement to the Fine aggregates and Cement respectively. The problems taking place in our environment that can be reduced by using the byproducts of the industries as an admixture is also discussed. CBA and fly ash taken in different proportions and also along with other materials like rice husk, super plasticizers, silica fumes and other such materials give rise to different mechanical and fresh properties impacting the strength and usability of the concrete. This papers also brings into account the flow ability of the concrete as in the case of self compacting concrete setting down of the mixture plays a vital role in the use of such concrete in the infrastructure world. The experiments and tests conducted by various authors are summarized and the results have been depicted in graphical format to show the comparison between them. The use of CBA and fly ash in the construction sector highly promotes the judicious use of such materials. Also that the safety and well being of the environment is the major concern and sustainability is one such pillar to rely upon, use of such byproducts can lead to a better environment for the generations to come. This paper can act as the gist for the probable use of the self compacting concrete in various proportions on the basis of the requirements of construction.

References

1. RafatSiddique, ParatibhaAggarwal ,Yogesh Aggarwal. Influence of water/powder ratio on strength properties of self-compacting concrete containing coal fly ash and bottom ash, Construction and Building Materials, Volume 29,April 2012,pp: 73-81
2. A.R. Esquinas, J.I. Álvarez, J.R. Jiménez, J.M. Fernández, Durability of self-compacting concrete made from non-conforming fly ash from coal-fired power plants, Construction and Building Materials, Volume 189, 20 November 2018, pp:993-1006.
3. Khongpermgoson, Penpichcha&Boonlao, Kamonrat&Ananthanet, Natthasing&Thitithananon, Thanaphon&Jaturapitakkul, Chai &Tangchirapat, Weerachart&Cheah, Chee. (2020). The mechanical properties and heat development behavior of high strength concrete containing high fineness coal bottom ash as a pozzolanic binder. Construction and Building Materials. 253. 119239. 10.1016/j.conbuildmat.2020.119239.
4. Promsawat P, Chatveera B, Sua-iam G, Makul N, Properties of self-compacting concrete prepared with ternary Portland cement-high volume fly ash-calcium carbonate blends, Case Studies in Construction Materials (2020), doi: https://doi.org/10.1016/j.cscm.2020.
5. F.A.Mustapha,A.Sulaiman,R.N.Mohamed,S.A.Umara.The effect of fly ash and silica fume on self-compacting high-performance concrete,SIE 2019: Sustainable & Integrated Engineering International Conference39(2), 2021, Pages 965-969.
6. Heba A. Mohamed. Effect of fly ash and silica fume on compressive strength of self-compacting concrete under different curing conditions, Ain Shams Engineering Journal, Volume 2,June 2011, PP: 79-86.
7. BinuSukumar, K. Nagamani, R. SrinivasaRaghavan. Evaluation of strength at early ages of self-compacting concrete with high volume fly ash, Construction and Building Materials 22 (2008) pp:1394–1401.
8. Malkit Singh, RafatSiddique. Strength properties and micro-structural properties of concrete containing coal bottom ash as partial replacement of fine aggregate, Construction and Building Materials,Volume 50, 2014a, PP: 246-256
9. Malkit Singh, RafatSiddique. Compressive strength, drying shrinkage and chemical resistance of concrete incorporating coal bottom ash as partial or total replacement of sand,Construction and Building Materials,Volume 68, 2014b, PP: 39-48
10. Malkit Singh, RafatSiddique. Effect of coal bottom ash as partial replacement of sand on workability and strength properties of concrete,Journal of Cleaner Production,, 2016, Volume 112, Part 1PP: 620-630
11. S C Bostanci. Coal ash use as a cement replacement in concrete production, 2020 IOP Conf. Ser.: Material Science Engineering.
12. Timothy Zhi Hong Ting, Muhammad EkhlasurRahman, Hieng Ho Lau. Sustainable lightweight self-compacting concrete using oil palm shell and fly ash,Construction and Building Materials, Volume 264, 2020,120590

13. RafatSiddique. Compressive strength, water absorption, sorptivity, abrasion resistance and permeability of self-compacting concrete containing coal bottom ash, Construction and Building Materials , 2013, vol 47 , pp: 1444-1450.
14. Muhammad HamzaHasnain, UsmanJaved, Ather Ali, Muhammad SaeedZafarEco-friendly utilization of rice husk ash and bagasse ash blend as partial sand replacement in self-compacting concrete,Construction and Building Materials,Volume2021,volume 273 pp:1-21.

IOP Conf. Series: Materials Science and Engineering 1236 (2022) 012003 doi:10.1088/1757-899X/1236/1/012003

Quantification of Non-conventional Brick's Characteristic Compressive Strength

Jyoti Prakash Giri[1] and Monalisa Priyadarshini[2]

[1]Assistant Professor, Department of Civil Engineering, GMR Institute of Technology, Rajam, Andhra Pradesh, India, 532127, Email: jyotiprakash.g@gmrit.edu.in.

[2]Assistant Professor, Department of Civil Engineering, Government College of Engineering Kalahandi, Bhawanipatna, Odisha 766002, India, Email: monalisa@gcekbpatna.ac.in.

Abstract. Now-a-days researches are concentrating on the utilization of waste materials in construction industry. Fly ash is one of the waste originated from industrial, is the main concern in the present study for brick manufacturing. As fly-ash is a by-product, its properties may vary from source to source resulting variation in strength of the brick. To confirming this, the main focus of the present study was to fabricate a hydraulically operated semi-automatic brick pressing machine followed by preparation of fly ash brick using that and study on the variation of brick strength. Later on, a statistical analysis is conducted based on the experimental compressive strength data for the fly ash bricks. The normal distribution curves are plotted followed by the estimating the constraints of normal distribution, which is the most common adopted distribution function in concrete. From the analysis, the characteristic compressive strength at 95% probability level is obtained as 3.46 and 5.42MPa respectively for the bricks (M1 and M2 type).

1. Introduction

Brick industry is one of the oldest one and popular for burnt block achieved after burning of clay in a kiln. As brick is a major construction material in the construction industry, it is also considerable that use of clay bricks creates a depletion of natural resources. Hence, numbers of researchers were explored the use of different waste materials to produce brick and try to conserve the natural clay. Generally, fly ash (FA) is a fine waste material obtained from power plant after coal combustion. This waste usually retains pozzolanic properties and offered advantages by enhancing the strength of concrete. This same pozzolanic behaviour when reacts with lime also found that helps in improvement of the quality of bricks compared to conventional clay brick [1]. The usage of fly ash in non-conventional bricks has been confirmed by different researchers and observed that, different source of fly ash offering different performance [1-8].Similarly compaction process of brick also plays an important role for their quality. Keeping all point in mind, the objective of the study is to fabricate a semi-automatic brick pressing machine followed by a probabilistic analysis on compressive strength of compacted bricks based on the experimental result at 95% level.

2. Semi-Automatic Brick Pressing Machine

As fabrication of brick pressing machine comprises a part of the present study, it was proposed to prepare the fly ash bricks imparting with other binding materials with desired pressure which resulting a dense compacted bricks for further observations in economical way. This machine marked as semi-automated, i.e. the functions of pressing and extraction from the brick mould are automated, whereas feeding of material into the machine is manual. The details of the pressing machine are summarized in Table 1and Figure.1.

Table 1. Detail specification of machine

Machine capacity	300 numbers of `bricks 8 hours
Mould size	190mm x 90mm x 90mm
Brick /stroke	2
Power of reciprocal type motor	2kW
Load	12-15 kN
Cylinder (Hydraulically operated)	*for pressing:* inner diameter of 125mm with outer diameter of 145mm, a shaft of diameter 60 mm
Table	cast iron with 850mm ×300×800mm size.
Hopper capacity	13590 cc
Die	200 x 200 x100 mm.
Hydraulic oil tank	45 lit. capacity
Mixer	Manual

Figure 1. Semi-automatic hydraulically operated brick pressing machine

3. Experimental Programme

3.1 Materials

The main aim of this experimental programme involves with the casting and testing the fly ash bricks. For this, the fly ash was collected from two different sources such as NALCO, Damonjodi and NTPC, Kaniha. The specific gravity of both the fly ash was found in laboratory as 2.28 and 2.19 respectively. The raw materials used to prepare these non-conventional bricks were fly ash, sludge lime, sand and gypsum based on the proportions suggested by NTPC report (Table 2). The sand was collected from a local river bed of Sagada. The fineness modulus and specific gravity of the sand was found as 2.37 and 2.65 in the laboratory respectively. The sludge lime was obtained from a nearby waste water treatment plant having specific gravity of 2.53 and the gypsum was collected from local source having specific gravity of 2.34.In the present study, two types of mixes were compacted using the fabricated pressing machine marked as M1(fly ash from NTPC, Kaniha) and M2 (fly ash from NALCO, Damonjodi). After compaction, the average strength at 21days for both the mixes were recorded to do the probabilistic analysis and to obtained the characteristic compressive strength at 95% confidence level.

Table 2. Details of the composition taken in bricks

Mix Id	Materials			
	Fly ash	Sand	Sludge Lime	Gypsum
M_1	45%	30%	20%	5%
M_2	45%	30%	20%	5%

3.2 Preparation of Bricks

First the materials were weighted (Table 2) as per the requirement and then mixed homogenously by adding required amount of water. After a homogenous mix it was fed into the hopper which is placed just above the brick mould. From the hopper the required quantity of materials was led into the mould and the mould was uniformly pressed hydraulically to formed dense bricks [Figure. 2(a)]. Then the moulds were hydraulically released and uplifted to release the pressed bricks. Using another hydraulic cylinder, the bricks were pushed to a ply board and allowed for an air curing for 1 day followed by 1 day sun dry. After 2 days, the same was transferred to water curing for 21 days. Figure.2 (b) and Figure.2 (c) represent the bricks before and after curing respectively.

Figure 2. (a) Compacted bricks from machine (b) Bricks before curing (c) Bricks after curing

4. Experimental Results

Fifty numbers bricks were casted using the machine for the each type mixes. The bricks after required curing were tested to observe the compressive strength using compression testing machine as shown inFigure.3.The average compressive strength of M1 and M2 fly ash brick samples are presented in Table 3.

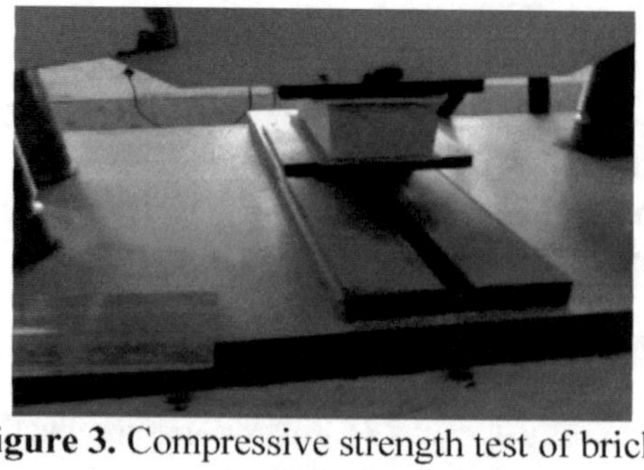

Figure 3. Compressive strength test of bricks

Table 3. Compressive strength of bricks

Mix Id	Average compressive strength (MPa)
M_1	3.22
M_2	5.18

5. Probabilistic Analysis

The real-world engineering situations are to mainly minimize with the uncertainty through an appropriate probabilistic model [9]. The normal distribution (Eq.1) which is the most commonly used distribution in statistics and it also applicable to a wide range of events in Civil Engineering applications [6-7].

$$f(x)=1/(\sigma\sqrt{2\pi})\exp[-1/2\{((x-\mu))/\sigma\}^2] \quad (1)$$

Where μ and σ are represented as the mean and standard deviation

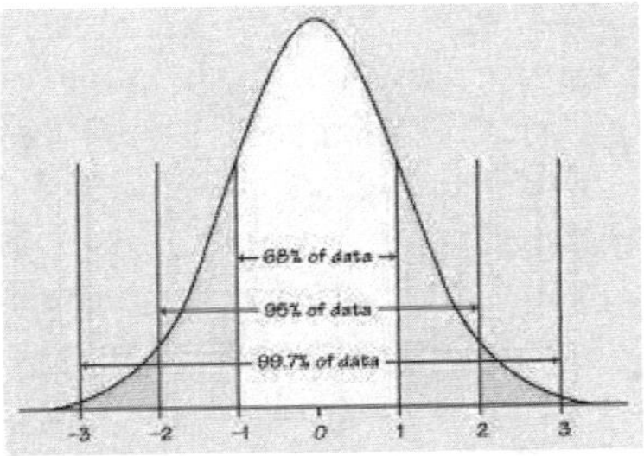

Figure 4. Normal distribution curve

The probability is obtaining randomly and it is a score to be evaluated from a normal distribution using an empirical rule. From Figure.4, it can be observed that 95% of values fall within the two standard deviations from the mean. The compressive strength of hundred numbers of non-conventional fly ash bricks after 21 days were the initial data for the normal probabilistic analysis. From this compressive strength, the estimated constraints were evaluated for normal distribution summarized in Table 4. By considering these estimated constraints, the normal distribution curves for brick M_1 and M_2 was presented in Figure.5. From the Figure.6, the characteristic compressive strength value at 95% probability level for mixes M_1 and M_2 bricks were estimated and shown in Table 5.

Table 4. Estimated constraints of normal distribution

Estimated Parameters	M_1	M_2
Minimum value	1.97	2.73
Maximum value	5.86	8.22
Mean	3.22	5.18
Standard deviation	0.52	0.58
Variance	0.30	0.47
Skewness	0.00	0.00
Kurtosis	0.00	0.00

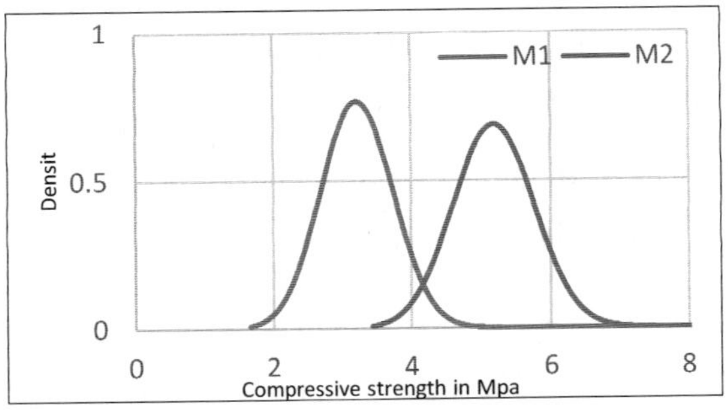

Figure 5. Normal distribution plot for the bricks (M_1 and M_2)

Table 5. Characteristic compressive strength of bricks at 95% probability level

Mix Id	Mean (μ)	Standard deviation (σ)	Characteristic Compressive Strength (MPa)
M_1	3.22	0.52	3.46
M_2	5.18	0.58	5.42

6. Conclusions

Based on the above experimental and probability analysis, the non-conventional bricks compacted with fly ash, sand, sludge lime and gypsum, the highest compressive strength was observed during the experimental work for the brick (M_2)prepared with fly ash collected from NALCO. From the statistical analysis evaluated for the normal probability distribution at 95% of confidence level were observed highest strength for the same M_2 mix. The reason behind the highest characteristic compressive strength for M2 mix may be due to the aluminum content available in the fly ash as it was obtained from NALCO whereas another one from coal. However, both the mixes were offered acceptable result (> 3.5MPa) according to codal provision [10]. From this it can be concluded that the strength behavior of bricks may vary with the source of fly ash. Addition to this, all source of fly ash used in brick preparation are confirming their performance requirement as per Indian specification which helps to conserve the natural resources as well as protecting the environment.

7. References

1. Kumar R, Deepankar, A and Najia L 2015 Properties of Non-Conventional (Fly Ash) Brick: An Experimental Study *International Journal of Engineering Trends and Technology* 24, pp 198-204.
2. Kumar S 2002 A perspective study on fly ash–lime–gypsum bricks and hollow blocks for low cost housing development *Construction and Building Material* 16, 519–525.
3. Kumar V, Mathur M and Kharia P S 2003 Fly ash management: Vision for the New Millenium *Technology Information Forecasting and Assessment Council News and Views* 25, March.
4. Kumar R and Hooda N 2014 An Experimental Study on Properties of Fly Ash Bricks *International Journal of Research in Aeronautical and Mechanical Engineering,* pp 56-67.
5. Zhang L 2013 Production of Bricks from Waste Materials-A Review *Construction and Building Materials* 47, pp 643-655.
6. Priyadarshini M, Patnaik M and Giri J P 2018 A probabilistic approach for identification of compressive strength of fly ash bricks *Innovative Infrastructure Solutions* 3(1), 56.
7. Priyadarshini M, Giri J P and Patnaik M 2021 Variability in the compressive strength of non-conventional bricks containing agro and industrial waste *Case Studies in Construction Materials* 14, e00506.
8. Alaa S and Ali A M 2013 Manufacturing of Bricks in the Past, in the Present and in the Future: A state of the Art Review *International Journal of Advances in Applied Sciences 2*, pp 145-156.

9. Ang A H S and Tang W H 2007 Probability concepts in Engineering Emphasis on applications to civil and environmental engineering, 2nd Edn. Wiley, New York (2007).
10. IS 12894-2002 Pulverized fly ash-lime bricks- Specification *Bureau of Indian Standards* New Delhi, India.

Response of Different RC Slab Systems in Buildings to Seismic Excitations

Ramakant Azad[1], Saraswati Setia[2]

Civil Engineering Department, National Institute of Technology Kurukshetra

Kurukshetra

ramakantazad025@gmail.com

Abstract. In the current study, the response of flat, grid and conventional slab against seismic excitation is studied and comparative analysis is performed to establish which among the three slab system would be most effective and safe in an earthquake scenario. The parameter such as storey displacement, drift, shear, and base shear of each slab arrangement is extracted from ETABS which is general structural design software for comparison. In the analysis, three different building specimen with the aforementioned slab type having length, width and height of 40m, 30m and 30m respectively are designed as per Indian design code IS: 456:2000 and analyzed as per IS:1893:2016.

Keywords: Storey displacement, Seismic force, Storey drift, Storey shear, Base shear, ETABS.

1. Introduction

The population of the world is increasing day by day. Besides this increase in population, housing requirements are also increasing [1]. Urbanization is the displacement of people from rural areas to urban areas which helps to decrease the proportion of the rural population. Nowadays, people are facing problems due to land scarcity and the cost of land. Therefore, there is required for vertical development of the building.[2]. Due to urbanization, the construction of high-rise buildings is very frequent. But, the tall buildings are not properly constructed and designed for the resistance of earthquake loads [3]. In earlier days, only gravity loads are considered for the design of buildings, but nowadays both gravity and lateral loads are considered for the design of buildings. Before the actual erection of the high-rise buildings, it should be analytically or numerically analyzed for seismic loads. Earthquake load is the most important reason for the failure of many tall building structures worldwide [4].

The storey shear force, storey deflection, storey drift ratio and shear at base play a crucial role to judge the building against the earthquake load [2]. Storey displacement is defined as the lateral displacement of the storey level to the ground floor. The difference between the lateral displacement of the successive floor is known as storey drift and the ratio of storey drift to the height of that floor is called storey drift ratio [5]. Storey shear is defined as the lateral force acting on each storey due to lateral force. It is a lateral force maximum at the bottom storey of the building and minimum at the top of the building. The total lateral load acting at its base is known as the base shear. [6]. From the last twenty years, there is a substantial rise in the number of tall buildings and the development is towards these types of structures due to congestion in cities. The lateral loads are the primary concerns in these buildings.

1.1 Conventional Slab System

The conventional slab is directly sustained on columns and beams. The depth of the beam is large and the thickness of the conventional slab is small. The weight of the slab is moved from slab to beams and beams to columns. The conventional slab needs more formwork as compare to a flat slab.

1.2 Flat Slab System

The flat slab is directly supported by columns or column head. Generally, it doesn't have beams. A flat slab is also known as a beam-less slab. The slab Loads are directly transmitted to columns and the columns to the footing. The flat slab requires less formwork. In this type of slab system, a flat ceiling is obtained gives a good aesthetic appearance. The flat slab is easy to construct as compared to other slabs.

The rectangular portion provided below the slab and above the column in the drop panels. The contact surface area between the column and slab increased in the drop panel, which helps to distribute the load from slab to column. The shear strength and negative moment capacity increase in the drop panel. By stiffening the flat slabs, the drop panel decreases deflection [9].

1.3 Grid/ Waffle Slab System

A grid slab is a concrete structure in which beams are provided at square or rectangular intervals in the perpendicular direction with slabs. These slabs are used at the entrance of the auditorium, theatre, showrooms, restaurants, malls, for good aesthetic view and also used for lighting, where column-free space is required. The waffle slab is stronger as compared to other slabs [10].

A waffle slab is employed where large spans are essential to escape a lot of columns prying with space. It is generally suitable for greater than 12 m spans. A large number of loads are held by using waffle slabs compare to conventional concrete slabs.

1.4 Research Objectives

The main objectives of current work are summed as under:

1. To study the performance of (G+9) multistoried building for slab systems under the earthquake load condition i.e. 1.5(DL+EQx) and 1.5(DL+EQy) for earthquake zone IV as per IS 1893-2016 part-1: Criteria for Earthquake resistant structure.
2. To study the seismic behaviour of different slab systems for various responses like storey displacements, base shear, story drift ratio and storey shear, and using ETABs software.
3. To compare the earthquake response of (G+9) multistoried building for different slab systems under different seismic load conditions in terms of several aforementioned parameters.

2. Method of Analysis

2.1 Methodology

The analysis of different slab system which considered has been done by using ETABS software. There are mainly three parts for structure using ETABS software.

1. Modelling
2. Analyzing
3. Designing

IOP Conf. Series: Materials Science and Engineering 1236 (2022) 012004 doi:10.1088/1757-899X/1236/1/012004

1. In the modelling parts, first define the material properties, section properties, diaphragm, load pattern, load combinations, mass source, after that draw column, beam, slab, drop, etc., and in last part of modelling is assign the frame loads, shell load, diaphragm, restraints, and auto floor mess.
2. After the modelling part, analyze the check model to find errors. To check the model, no warning message will create, then go for run analysis.
3. After the analysis, we extract the followings results parameter like storey shear, storey drift, storey deflection, for comparing the behaviour of flat slab, waffle slab, and conventional slab arrangements.

2.2 Building Specification

Consider G+9 multistoried RC building with each floor height of 3m and has 5 bays of 8m in the longitudinal direction and 6m in the and transverse direction. The slab thickness of each floor is considered 250 mm. The dimension of the beam taken for ten stories is 300 x 250mm. The column size taken for ten stories is 900 x 900 mm.

Table.1

S.NO.	SPECIFICATIONS	TYPES OF SLAB SYSTEM		
		Conventional Slab	**Flat Slab**	**Waffle Slab**
1	Plan Dimension of building	40m × 30m	40m × 30m	40m × 30m
2	Grid Length in longitudinal direction	8m	8m	8m
3	Grid Length in transverse direction	6m	6m	6m
4	Total number of storey level	10	10	10
5	Height of each floor	3m	3m	3m
6	Depth of slab	0.25m	0.25m	0.25m
7	Cross-section of Beam	0.30m × 0.25m		0.30m × 0.25m
8	Cross-section of Column	0.90m × 0.90m	0.90m × 0.90 m	0.90m × 0.90m
9	Size of Ribs			125 mm × 125 mm
10	Spacing of Ribs			1400 mm
11	Grade of Concrete	M25	M25	M25
12	Grade of Steel	Fe-415	Fe-415	Fe-415

IOP Conf. Series: Materials Science and Engineering 1236 (2022) 012004 doi:10.1088/1757-899X/1236/1/012004

2.3 Loading Condition

During the analysis of the multistoried building, the loads are considered as

1. **Dead load**: It is calculated as per IS -875 (Part 1) : 1987
   a) Plate load =Concrete density× Slab thickness
   = 25 KN/m^3×0.25m = 6.25 KN/m^2
   b) Masonry load = 1 KN/m^2
   c) Floor finishing load = 1.5 KN/m^2
   Total weight of slab = 8.75 KN/m^2
2. **Seismic load**: It is taken as according to IS-1893 (part 1): 2016
   Seismic zone: IV (Z=0.24)
   Response reduction factor: 5
   Importance Factor: 1.5 (Very Important Building)
   Rock and Soil Site Factor: 1 (Medium Soil Building)
   Type of Structure: 1
   Damping of structure: 5% (0.05)
   Type of soil: II (Medium soil)

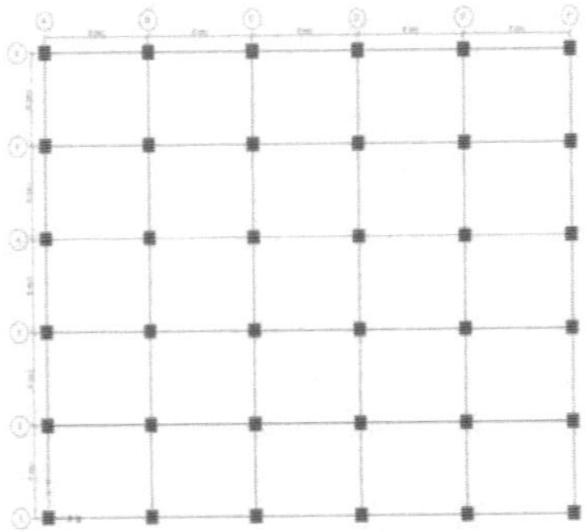

Fig.1 Plan of G+9 models of Conventional slab

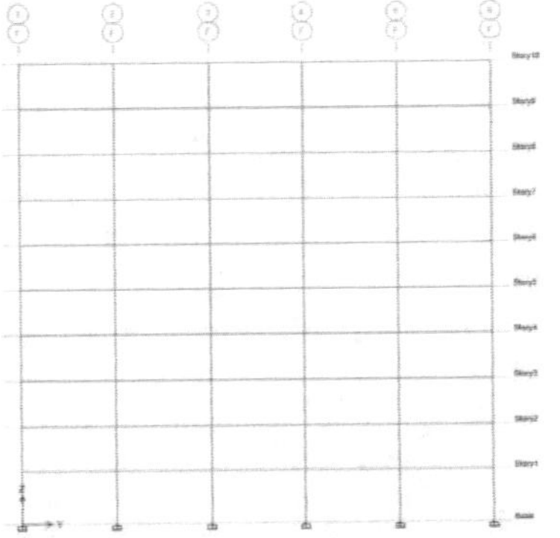

Fig.2 Elevation of (G+9) model Conventional slab

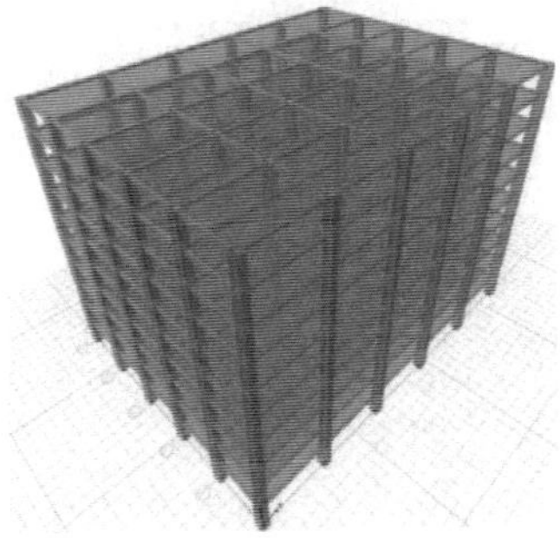

Fig.3 3-Dimensional view of Conventional Slab System

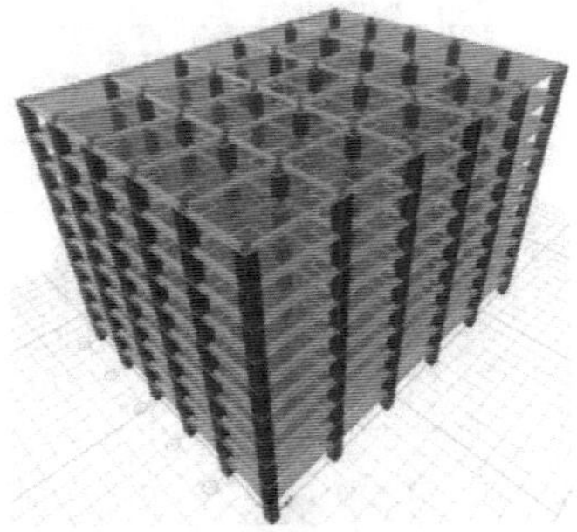

Fig.4 3-Dimensional view of Flat Slab System

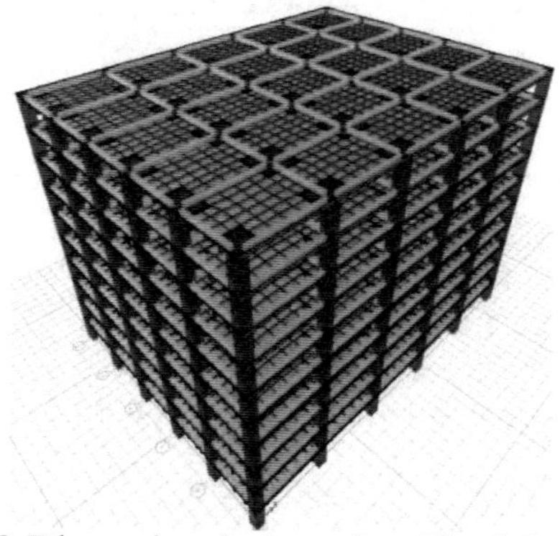

Fig.5 3-Dimensional view of waffle slab system

3. Results and Discussions

The results are provided in the form of a chart in the subsequent figures after analyzing all the three buildings using the software. The earthquake load patterns of 1.5(DL+EQ_x) and 1.5(DL+EQ_y) are considered.

3.1 Effect of storey displacement for different slab system

Maximum storey displacement is calculated for the two load patterns are 1.5(DL+EQ_x) and 1.5(DL+EQ_y) in longitudinal and transverse direction correspondingly. In fig.3.1 and fig.3.2, the abscissa denotes the storey level, and the ordinate denotes the maximum displacement. It can be concluded that the displacement is increasing from the ground level to the top storey level of the building.

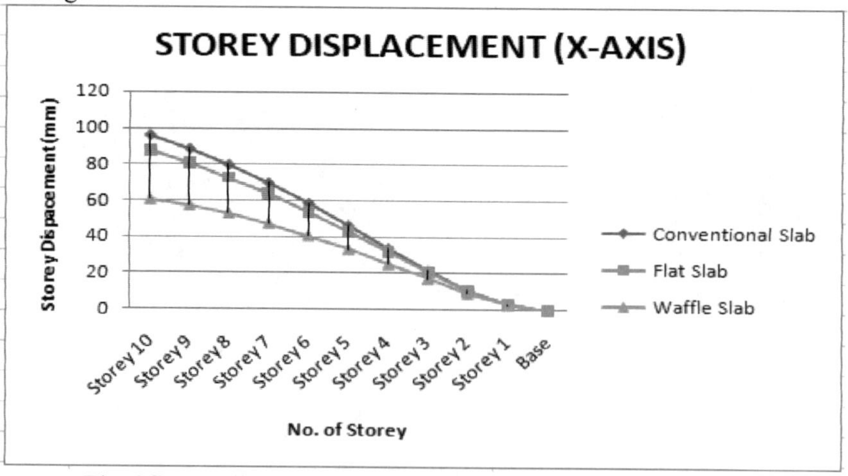

Fig.6 Storey displacement for 1.5(DL +EQ_x) in X-axis direction

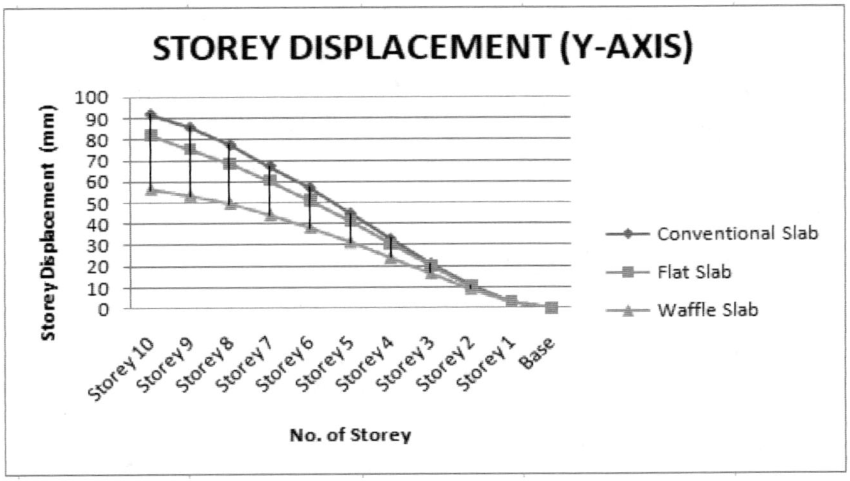

Fig.7 Storey displacement for 1.5(DL + EQ_y) in Y-axis direction

3.2 Effect of Storey Drift for different slab system

Maximum storey drift is finding for the load patterns are 1.5(DL+EQx) and 1.5(DL+EQy) in longitudinal and transverse direction correspondingly. In fig.3.3 and fig.3.4, the abscissa denotes the storey level, and the ordinate denotes the maximum drift. It can be concluded that the Storey drift is increasing first and after that decreasing from bottom to top of the building. The storey drift is observed highest at the fifth storey for each slab system.

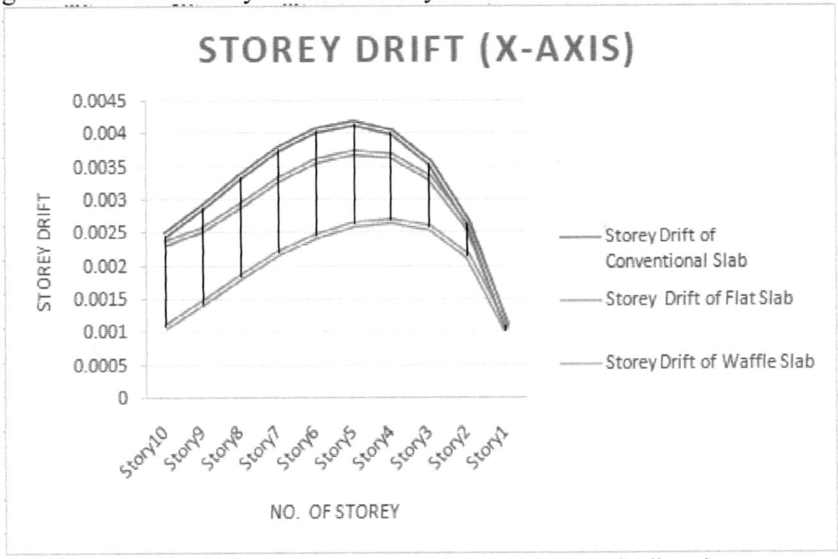

Fig.8 Storey Drift for 1.5(DL+EQ_x) in X-axis direction

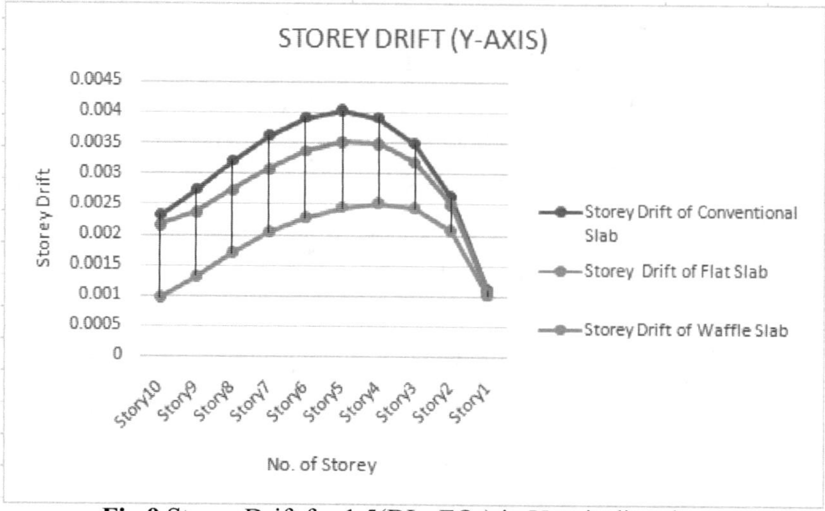

Fig.9 Storey Drift for 1.5(DL+EQ_y) in Y-axis direction

3.3 Effect of Storey Shear

Maximum storey shear is calculated for the two load patterns are 1.5(DL+EQx) and 1.5(DL+EQy) in longitudinal and transverse direction correspondingly. In fig.3.5 and fig.3.6, the abscissa denotes the storey level, and the ordinate denotes the maximum storey shear. It can be concluded that the Storey shear is decreasing from bottom to top of the building. The storey shear force is decreasing from the ground level to the top storey level of the building.

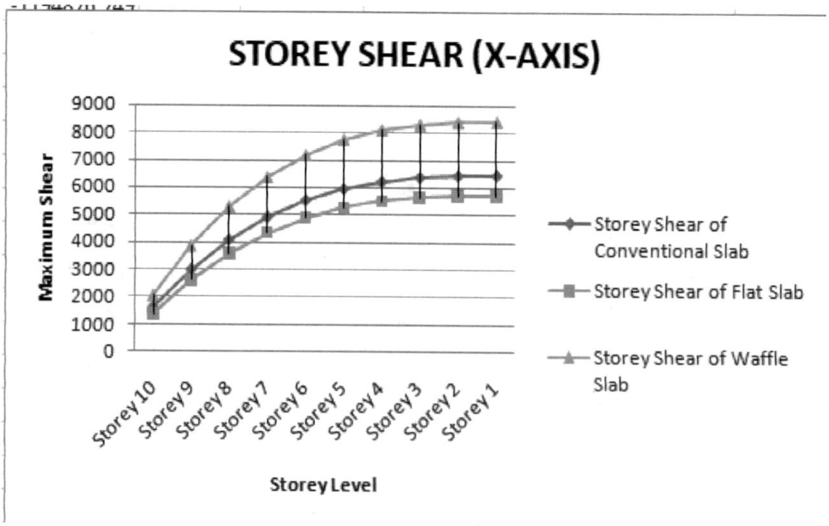

Fig.10 Storey Shear for 1.5(DL+EQ_x) in X-axis direction

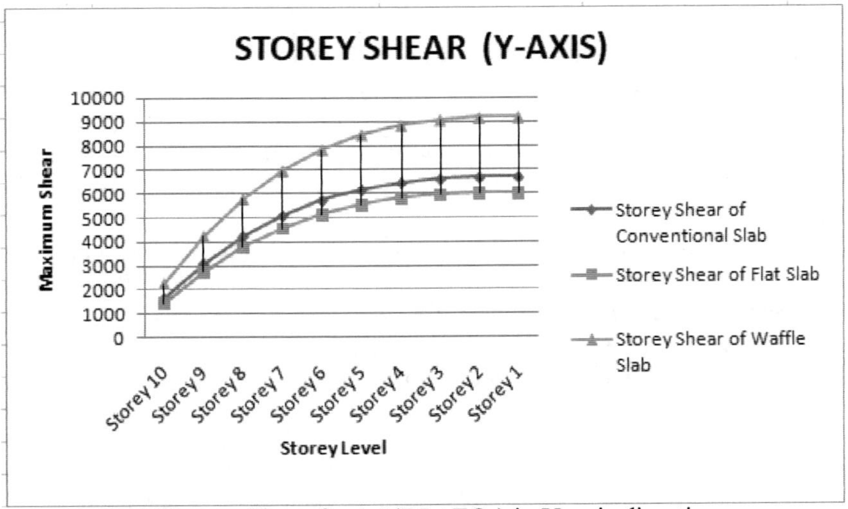

Fig.11 Storey Shear for 1.5(DL+EQ_y) in Y-axis direction

3.4 Effect of Base Shear

In fig.3.7, the abscissa denotes the types of the slab, and the ordinate denotes the base shear. It clearly shows that the base shear is higher for waffle slab arrangement compare to flat slab and conventional slab arrangement.

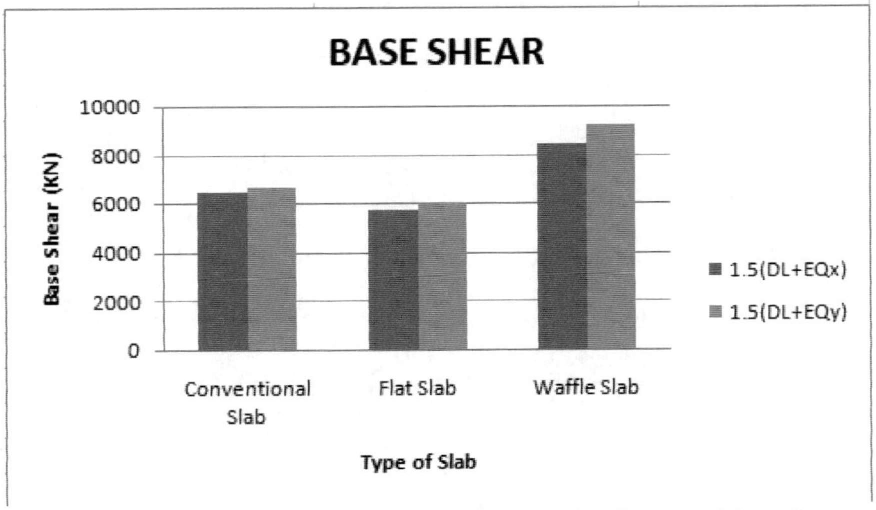

Fig.12 Base shear for different slab under loading condition of 1.5(DL+EQ_x) and 1.5(DL +EQ_y)

4. Conclusions

- The storey deflection is minimum for waffle slab and maximum for conventional slab for the similar dimensional area of the building. The story displacement is rising with a rise in height of the storey.
- The maximum storey deflection for the conventional slab arrangement is 44.44% higher than the waffle slab arrangement and 58.81 % higher than the flat slab arrangement in the longitudinal direction.
- The storey drift is minimum for waffle slab arrangement and maximum for conventional slab arrangement for the similar plan area of the building. The storey drift ratio is increasing first and after decreasing with an increase in storey height. The maximum storey drift is found at the fifth storey of the building for considered loading condition for the slab system.
- The maximum storey drift for a conventional slab is 11.56% more than the flat slab and 57.2% more than the waffle slab system in the longitudinal direction.
- The storey shear force is maximum for waffle slab arrangement and minimum for flat slab arrangement.
- The maximum storey shear force for a waffle slab system is 47.4% more than the flat slab system and 31% more than the conventional slab system in the longitudinal direction.

References

1. Kumar, C.J.D., Venkat, L: Analysis of multi-story building with precast load-bearing walls, International Journal of Civil and Structural Engineering (2013) Volume 4, No 2
2. Yadav, J.C., Reddy, R.C.: Dynamic analysis of G + 20 residential building in zone2 and zone5 by using ETABs, Volume VIII, Issue 3, IJPES-(2017), pages 333-346
3. Kumar, P.E., Naresh, A.: Earthquake Analysis of Multi Storied Residential Building - A Case Study, IJERA-(2014) Vol. 4, Issue 11, pages 59-64
4. Baikerikar, A., Kanagali, K.: Study of lateral load resisting systems of variable heights in all soil types of high seismic zone, Volume: 03, Issue: 10 IJERT -(2015), pages109-119.
5. Mohana H.S1, Kavan, M.R.: Comparative Study of Flat Slab and Conventional Slab Structure Using ETABS for Different Earthquake Zones of India, Volume: 02, Issue: 03, IRJET-(2015), pages1931-1936
6. Navale, B., Singh, P., Wakechaure, M.R.: Comparative study of flat slab & conventional slab structure using software (2019)
7. Naresh, K.: Comparative study of flat slab and conventional slab, International Research Journal of Engineering and Technology (IRJET) (2019)
8. Bhaduria, S.S.Chhugani, N.: Comparative Analysis and Design of Flat and Grid Slab System with Conventional Slab System, Volume: 04, Issue: 08, IRJET-(2017), pages 2314-2329
9. Aman, S.A., Khaliq, M.: Analysis and Design of a Multi Storey Building with Flat Slab (C+G+9) Using ETABS, Volume.06, Issue.01, IJITECH- (2018), Pages:0176-0183
10. Santhosh, C., Wadki, V.: Analysis and Design of Multistory Building with Grid Slab Using ETABS, Vol. 5, Issue 9, IJIRSET-(2016), pages 17201-17208

IOP Conf. Series: Materials Science and Engineering 1236 (2022) 012005 doi:10.1088/1757-899X/1236/1/012005

Performance Based Seismic Design of Steel Structure Using Non-Linear Dynamic Analysis

Choimei Moileen Semdok and Babita Saini
Department of civil Engineering, National Institute of Technology, Kurukshetra, 136119, India
csemdor_31902301@nitkkr.ac.in
bsaini@nitkkr.ac.in

Abstract. In the recent decades, it was observed that the performance of buildings that were designed using the traditional building codes were not satisfactory under seismic effects. There was increased demand for buildings that would not only provide life safety under the design earthquake and prevent collapse under major earthquakes, but also include other performance objectives such as the time taken for the building to be occupied post-earthquake, cost of repair and so on. This resulted in the research and development of a performance based seismic design that would incorporate these demands in the design. This study presents a performance based seismic design of a 5-storey steel special moment resisting frame building by using non-linear dynamic analysis to meet multiple performance objectives. The seismic performance was evaluated based on the confidence level, which indicates the ability of the building to meet the target performance objective, determined for the inter-storey drift parameter, using the guidelines given in FEMA-350. The results show that the designed building satisfies the minimum confidence level requirement for each performance objective. The study concludes that performance based seismic design proves to be a reliable method for design of buildings that provide enhanced seismic performance.

Keywords: Performance based seismic design, Non-linear dynamic analysis, Bi-directional earthquake

1. Introduction

Several earthquakes that took place over the past few decades namely the Northridge earthquake which took place in the US in 1994 and the Kobe earthquake that took place in Japan in 1995, brought to surface the inability of the current building codes in predicting the behavior of buildings under seismic effects [1]. Buildings designed using the building codes are expected to provide life safety under design earthquake, without damage under minor earthquakes and prevent collapse under severe earthquakes. This however does not guarantee that the building would meet each of those criteria when subjected to seismic effects as the code does not explicitly include design for multiple limit states. Although buildings do survive the earthquakes, the economic losses because of the earthquakes, are unprecedented. FEMA 424 however mentions that this does not qualify buildings designed and built using the building codes as unsafe, instead the requirements in the building codes should be taken only as a minimal requirement for protecting public safety, and general welfare. There may be times when it will be beneficial or required to go beyond minimum safety requirements prescribed by codes [2]. Owing to these requirements, research for a more efficient design method that would tackle the shortcomings of the building codes and also incorporate the cost of damages, began, leading to the development of Performance Based Seismic Design (PBSD) procedure.

Several PBSD approaches were proposed such as Yield Point Spectra Method (Aschhen and Black,2000), Modified Lateral force procedure (Englekirk,200; Panagiotou and Restrepo,2007), Deformation Controlled design (Panaglotakoes and Fardis M.N, 1999), Direct Displacement Based Design (DDBD) (Priestly et al., 2003-2007) and Performance Based Plastic Design (PBPD) (Goel and Chopra 2007) [3]. Among these methods, the DDBD and PBPD are the most developed methods, although they are still being improved. These PBSD approaches aim to provide feasible steps for performance-based design of buildings by incorporating maximum design features while also avoiding the need for iteration for design.

Several studies were carried out in an effort to compare between PBSD and Force Based Design (FBD) methods to weigh the pros and cons of PBSD against FBD in terms of reliability, feasibility, economy, etc. Muljati et al [4] concluded that DDBD was superior in comparison to FBD in predicting the seismic demand of structure. Sutariya and Shah [5], Gamit and Amin [6] and Singh and Raju [7], all concluded that DDBD proved to be more economical than FBD. Vivinkumar R. V. and Karthiga S. [8] found that DDBD structure showed better performance as compared to FBD structures. Dubal et al [9] observed that PBSD method significantly enhanced the performance point of frames as compared to FBD method for all cases.

Studies have also been done on PBSD of various types of buildings configurations, height, lateral force resisting systems etc. Fox et al [10] designed a 7-storey building in New Zealand, with RC coupled walls using PBSD. Karthiga et al [11] compared FBD and DDBD of two-dimensional RC frames having mass irregularity as well as geometric irregularity in vertical direction. Dubal et al [9] analysed a 10-storey RC moment resisting frame having vertical geometric irregularity and soft storey using PBSD as well as FBD. Khedkar et al [12] studied the RC moment resistant frame having vertical setback using PBSD and FBD method as per IS 1893: 2002. Ccahuana et al [13] examined the seismic responses and damping parameters of a dual concrete building and a CBF frame using DDBD procedure to compare their base load and displacements.

Besides RC structures, various studies were conducted on steel structures as well. Lee et al [14] proposed a PBSD procedure to design SMRFs by using pre-selected yield mechanism and target drift and concluded that the proposed method was able to produce structures that would meet the pre-selected performance objectives. Zhang et al [15] put forward a PBPD methodology, which is a PBSD method, for designing high rise steel frames and concluded that PBPD frames displayed excellent seismic performance. T. J. Sullivan [16] formulated a DDBD procedure for eccentrically braced frames (EBFs) and concluded that the proposed methodology was promising. Malekpour et al [17] assessed the DDBD procedure for regular steel moment resisting frames and concluded that DDBD can be considered as an alternative to the current FBD methods. Wijesundara and Rajeev [18] proposed a DDBD procedure for steel CBFs by utilizing the yield displacement profile and concluded that the linear displacement shape proposed by Priestley et al [19] for low MRFs was substantially valid for low-rise CBF structures.

Although substantial efforts have been made towards PBSD of steel buildings, there are still lesser research works available as compared to reinforced concrete structures. Therefore, this study was conducted to further understand the behaviour of steel structures designed using PBSD. The performance of buildings considering multiple limit states (e.g., collapse, immediate occupancy, life safety) and enhanced performance objectives also needs to be investigated more and hence is included in this study. Furthermore, a 3D model of the building has been considered instead of the 2D models generally used, to better incorporate the multi-directional effects during seismic activity and the building is subjected to bi-directional earthquake effects by applying orthogonal pair of ground motions to the building, simultaneously in both directions instead of the generally used unidirectional earthquakes.

2. Overview and Design of the Building

A 5-storey special steel moment resisting frame (SMRF) office building has been designed for the study. The building has been assumed to be constructed in a region of high seismicity. The resistance to lateral seismic forces was provided by Seismic Force Resisting System (SFRS) consisting of special SMRFs arranged symmetrically at the perimeter in both the orthogonal directions, having three bays in E-W direction and four bays in N-S direction. The detailed arrangement of the SFRS has been given in figure1.

The building considered was rectangular in plan with five bays of 28ft in E-W direction and four bays of 23ft in N-S direction, giving a total floor area of 140x92 ft2. A storey height of 13ft was provided for all the levels. The building schematics has been shown in figure1 and the typical floor plan in figure 2.

Design Loads. The dead load from self-weight of the steel frame system and concrete deck of 4.8 inch of 4000 psi was automatically calculated by the software and included in the analysis. Super imposed dead load of 650 lb/ft from façade was provided at the perimeter of each floor, except on the roof where load of 165 lb/ft from parapet has been provided at the perimeter. A finishing load of 12 lb/ft2 was provided on all the

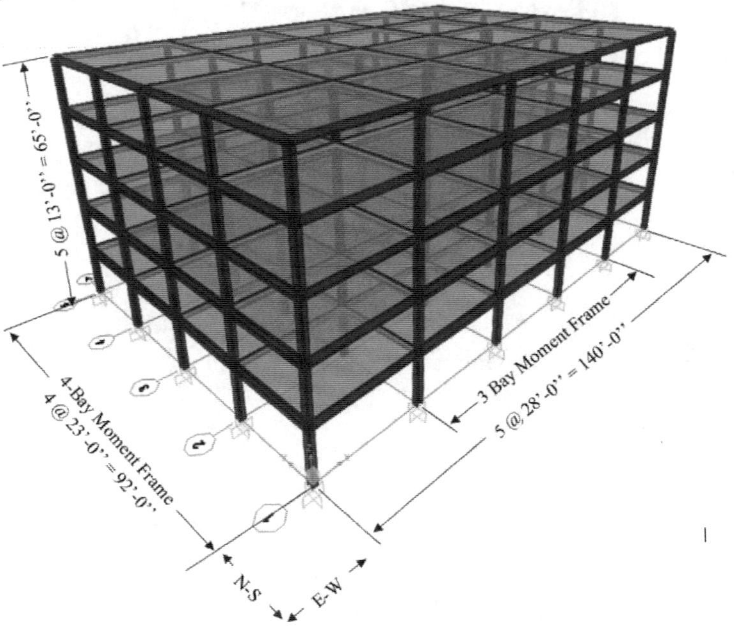

Fig. 1. Isometric View of 5-Storey building

floors and load coming from the equipment and miscellaneous loads assuming to be 15 lb/ft2 was also applied on all floors. Partition wall of 250 lb/ft is applied on all the floors

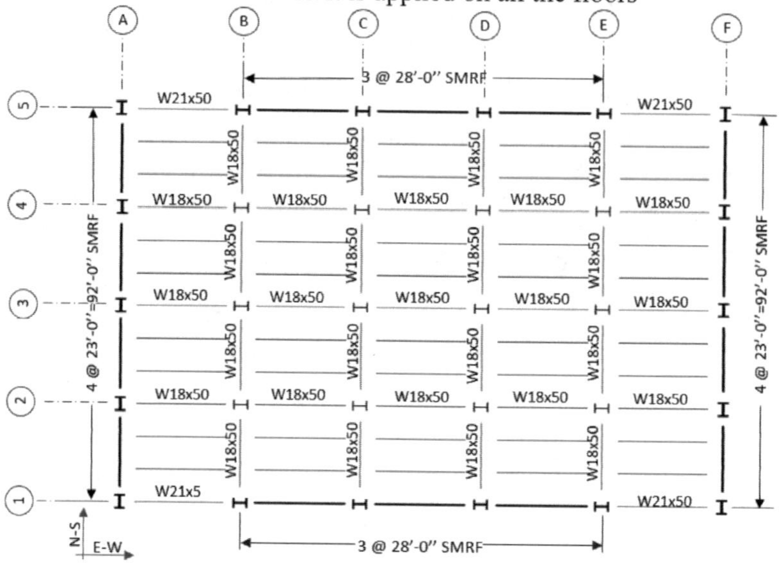

Fig. 2. Typical Floor Framing Plan

except the roof. Live load of 60 lb/ft2 has been provided on all the floors with a live load of 30 lb/ft2 on the roof.

Seismic Parameters. The SH provided in ASCE 7 is based on a risk-targeted design philosophy. It is defined as ground motions having 1% probability of causing total or partial structural collapse of an appropriately designed structure in 50 years. This ground motion intensity is designated in the code as MCER [20]. The SH used for the design have been given below.

- Building Risk Category: II ASCE 7 Table 1.5-1
- Soil site class: D, Stiff soil ASCE 7 Table 20.3-1
- Spectral response acceleration parameters:
  $S_s = 1.5g$ $S_1 = 0.67g$
  $S_{DS} = 1g$ $S_{D1} = 0.6g$
- Seismic Design Category: D ASCE 7 Table 11.6-1
  D ASCE 7 Table 11.6-2

A preliminary design was processed using Equivalent Lateral Force method as per ASCE 7-16 in SAP2000. The building was then analysed using Non-Linear Dynamic Procedure in order to enhance the design to meet the acceptance criteria. The building's performance has been evaluated as per FEMA-350. The guideline provides a formula for determining a level of confidence in the ability of the building to meet the desired performance objectives. FEMA-350 mentions three parameters, namely

IOP Conf. Series: Materials Science and Engineering 1236 (2022) 012005 doi:10.1088/1757-899X/1236/1/012005

inter-storey drift ratio (IDR), column axial load and column splice tension that require evaluation of confidence. However, in this current report, the building was designed as per the confidence level determined by only using the IDR parameter.

Ground Motion Data. A set of 12 ground motion records were selected comprising of

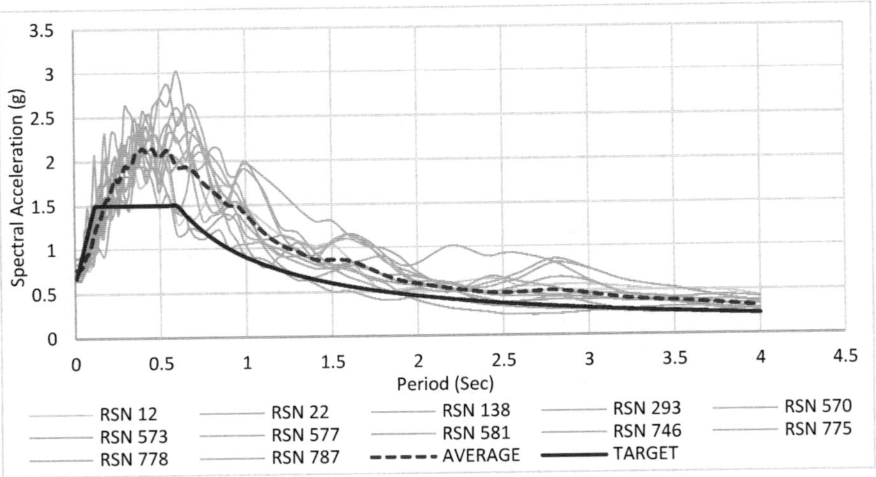

Fig. 3. BSE-2 Response Spectra

two horizontal components for each record. The ground motion characteristics have been provided in table1.

A graph showing the target response spectrum for BSE-2 Seismic Hazard Level (SHL) along with the response spectra of the selected ground motions and their corresponding average spectrum has been shown in figure 3.

The performance objective selected is Basic Safety Objective characterized by Collapse Prevention (CP) for Risk Targeted Maximum Considered Earthquake (MCER) with probability of exceedance (POE) of 2% in 50 years, Life Safety (LS) for Design Basis Earthquake (DBE) with POE of 10% in 50 years and Immediate Occupancy (IO) for Serviceability Earthquake (SE) with POE of 20% in 50 years.

IOP Conf. Series: Materials Science and Engineering 1236 (2022) 012005 doi:10.1088/1757-899X/1236/1/012005

Table 1. Ground motion data

No.	RSN	Event Name	Station Name	Scale Factor	Magnitude	Year
1	12	Kern County	LA-Hollywood Stor FF	9.46	7.36	1952
2	22	El Alamo	El Centro Array #9	12.56	6.8	1956
3	138	Tabas_Iran	Boshrooyeh	5.26	7.35	1978
4	293	Irpinia_Italy-01	Torre Del Greco	11.17	6.9	1980
5	570	Taiwan SMART1(45)	SMART1 C00	3.32	7.3	1986
6	573	Taiwan SMART1(45)	SMART1 I01	3.33	7.3	1986
7	577	Taiwan SMART1(45)	SMART1 O01	3.53	7.3	1986
8	581	Taiwan SMART1(45)	SMART1 O07	4.01	7.3	1986
9	746	Loma Prieta	Bear Valley #5_Callens Ranch	7.84	6.93	1989
10	775	Loma Prieta	Hollister-SAGO Vault	11.29	6.93	1989
11	778	Loma Prieta	Hollister Differential Array	2.17	6.93	1989
12	787	Loma Prieta	Palo Alto-SLAC Lab	2.65	6.93	1989

3. Results and Discussion

The results presented here are the average of the maximum response from all the 12 ground motion sets.

3.1 Inter-storey Drift Ratio

A summary of the mean IDRs for all the performance objectives have been shown in Table2. Drift limits as per FEMA-273 are also provided. However, it should be noted that these drift limits were prescribed for steel moment resisting frames without consideration of the gravity frames.

Table 2. Maximum Inter-storey drift ratio

Performance Level	X	Y	Drift Limit
CP	0.033	0.032	0.050
LS	0.025	0.024	0.025
IO	0.014	0.014	0.007

For CP, the IDRs were almost similar in both the orthogonal directions with maximum IDR of 0.033 and 0.032 occurring at the 3rd storey in E-W and N-S directions respectively. The maximum drift ratio limit of 0.05 given in FEMA-273 has been satisfied i.e., the drift ratio obtained was much lesser than the maximum limit of 0.05.

In the case of LS, the maximum drift limit provided by FEMA-273 is 0.025 for steel moment frames to be within the LS performance level. The result from the analysis showed the maximum IDR to be 0.025 and 0.024 for E-W and N-S direction respectively. The IDRs were at the edge of the drift ratio limit for LS performance level unlike the conservative result shown in the CP performance level.

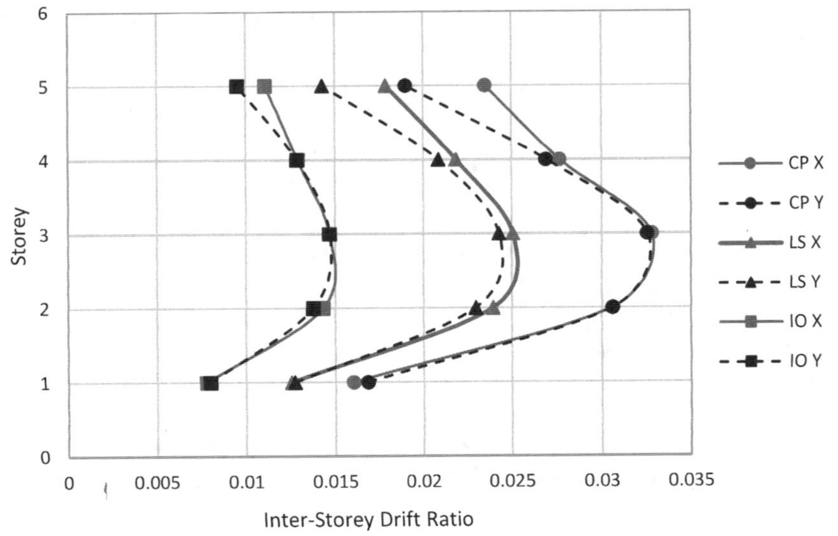

Fig. 4. Summary of Mean Inter-Storey Drift Ratio

For IO performance level, FEMA-273 specifies a drift limit of 0.007 for steel moment frames. The result from analysis showed the maximum IDR 0.014 for frames in E-W and N-S directions. This showed that the drift ratios obtained for IO performance level was twice as much from the drift ratio limit specified by FEMA-273 and hence did not satisfy the criteria.
A graph showing the summary of the mean of maximum IDRs for all performance levels has been shown in figure 4.

3.2 Confidence Level

The confidence level obtained and the minimum confidence level requirement as per FEMA-350 for each performance level has been provided in table 3. The acceptance criteria for LS are not provided in FEMA-350. However, for this study, the values were interpolated and evaluated for verification (as was suggested in FEMA-350 commentary).

Table 3: Confidence Levels

Performance Level	X	Y	Minimum Requirec
CP	>>99%	>>99%	90%
LS	97.77%	98.38%	70%
IO	52%	52%	50%

From table 3, it has been observed that the minimum confidence level requirement as per FEMA-350 was satisfied for all the performance levels.
Based on the acceptance criteria for IDR parameter as per FEMA-273 and FEMA-350, the results obtained showed conservative behaviour for CP performance level. The results obtained for LS just reached the drift limit provided by FEMA-273 but was conservative as per FEMA-350. However, the minimum confidence level for LS has been interpolated for this study and hence may not be accurate. In case of IO performance level, the IDR obtained did not satisfy the IDR limit given by FEMA-273 and it only nearly satisfied the minimum confidence level requirement provided by FEMA-350. Hence it can be seen that the ability of the building to satisfy the IO performance objective posed the highest risk and so it proved to be the governing criteria for design in this study.

4. Conclusion

This study has provided a PBSD of a steel SMRF office building with its seismic performance evaluated based on IDR parameter by using a reliability-based acceptance criteria in the form of confidence levels given by FEMA-350. The design has been verified successfully with the building's performance meeting the minimum confidence level requirements. The use of NDP, though tedious, has been proven to be the most accurate method and hence used in many experimental studies for establishing the benchmark of structure's response.
In conclusion, performance based seismic design has been successfully evaluated in this study and has proven to be a reliable method for design, providing enhanced seismic performance at multiple limit states.

References

1. US Geological Survey Publications Warehouse, https://pubs.usgs.gov/of/1996/ofr-96-0263/introduc.htm, last accessed 2021/06/23.
2. Federal Emergency Management Agency, FEMA P-424, Design guide for improving school safety in earthquakes, floods and high winds, December (2010).
3. Qammer, S. S., Dalal, P. S.: Development of Performance Based Seismic Design: A Review. INTERNATIONAL CONFERENCE ON ENGINEERING AND TECHNOLOGY, Coimbatore (2016).
4. Muljati, I., Asisi, F., Willyanto, K.: Performance of Force Based Design Versus Direct Displacement Based Design in Predicting Seismic Demands of Regular Concrete Special Moment Resisting Frames. THE 5TH INTERNATIONAL CONFERENCE OF EURO ASIA CIVIL ENGINEERING FORUM (EACEF-5), Procedia Engineering 125 1050 – 1056, Elsevier (2015).
5. Sutariya, K. N., Shah, A. B.: A Comparative Study of Force Based Design and Direct Displacement Based Design for R.C. Buildings. International Research Journal of Engineering and Technology, IRJET, Vol. 03, Issue 05 (2016).
6. Gamit, D. K., Amin, A. J.: Application of DDBD and FBD Methodology for 8-Story RC Frame Using IS 1893 Spectra. INTERNATIONAL CONFERENCE ON RESEARCH AND INNOVATIONS IN SCIENCE, ENGINEERING &TECHNOLOGY, ICRISET 2017, Vol. 1, pp. 69-79 (2017).
7. Singh, R. R., Raju, P. P.: Performance Based Seismic Design of RC Framed Building. International Journal of Civil Engineering and Technology, IJCIET, Vol. 8, Issue 4, pp. 401–412, April (2017).
8. Vivinkumar, R.V., Karthiga, S.: A Comparative Study on Force Based Design and Direct Displacement Based Design of Reinforced Concrete Frames. IJRET: International Journal of Research in Engineering and Technology, Vol. 04 Issue 04 (2015).
9. Dubal, R., Neha, G., Patil, G. R., Vasanwala, S., Modhera, C.: Application of Performance Based Seismic Design Method to Reinforced Concrete Moment Resistant Frame with Vertical Geometric Irregularity with Soft Storey. American Journal of Engineering Research, AJER, Vol. 03, Issue 12, pp. 54-61 (2014).
10. Fox, J. M., Sullivan, J. T., Beyer, K.: Comparison of Force-Based and Displacement Based Design Approaches for RC Coupled Walls in New Zealand. Bulletin of The New Zealand Society for Earthquake Engineering, Vol. 47, No. 3, pp. 190-205 (2014).
11. Karthiga, S., Tamilselvan K., Pavithra, C.: Comparison of Force Based Design and Direct Displacement Based Design of Two Dimensional Reinforced Concrete Frames with Vertical Geometric and Mass Irregularities. Journal of Industrial Pollution Control, ISSN 0970-2083 (2017).

12. Khedkar, S. A., Dubal, A. R., Vasanwala, A. S.: Performance Based Seismic Design of Reinforced Concrete Moment Resistant Frame with Vertical Setback. International Journal of Engineering Research & Technology (IJERT), Vol. 3 Issue 2, Paper ID IJERTV3IS21114 (2014).
13. Ccahuana, W., Coronel, J., Soto, J.: Direct Displacement-Based Design (DDBD) applied to dual wall-frame buildings and steel concentric braced frames. INTERNATIONAL CONFERENCE ON CIVIL ENGINEERING AND ARCHITECTURE, ICCEA, IOP Conf. Series: Materials Science and Engineering, vol. 910, IOP Publishing Ltd, France (2020).
14. Lee, S. S., Goel, C. S., Chao, H. S.: Performance-Based Seismic Design of Steel Moment Frames using Target Drift and Yield Mechanism. 13TH WORLD CONFERENCE ON EARTHQUAKE ENGINEERING, Paper No. 266. Vancouver, Canada (2004).
15. Zhang, Q., Xiong, E., Liang, X., Miao, X.: Performance-based plastic design method of high-rise steel frames. JVE International Ltd. Journal of Vibroengineering. Vol.19, Issue 3, pp. 2003-2018 (2017).
16. Sullivan, T. J.: Formulation of a direct displacement-based design procedure for steel eccentrically braced frame structures. 15TH WORLD CONFERENCE ON EARTHQUAKE ENGINEERING, WCEE, Lisboa (2012).
17. Malekpour, S., Ghaffarzadeh, H., Dashti, F.: Direct Displacement Based Design of Regular Steel Moment Resisting Frames. THE TWELFTH EAST ASIA-PACIFIC CONFERENCE ON STRUCTURAL ENGINEERING AND CONSTRUCTION, Procedia Engineering 14, pp. 3354–3361 (2011).
18. Wijesundara, K. K., Nascimbene, R., Sullivan, T.J.: Equivalent viscous damping for steel concentrically braced frame structures. Bulletin of Earthquake Engineering, vol.9(5), pp. 1535-1558 (2011).
19. Priestley, M. J. N., Calvi, C. M., Kowalsky, M. J.: Displacement based seismic design of structures, IUSS Press, Pavia, ISBN: 978-88-6198-0006. pp. 721, Italy (2007).
20. National Institute of Standards and Technology, NIST.: Assessment of First Generation Performance-Based Seismic Design Methods for New Steel Buildings. Volume 1: Special Moment Frames. (2015).

IOP Conf. Series: Materials Science and Engineering 1236 (2022) 012006 doi:10.1088/1757-899X/1236/1/012006

Behavior of Different Configuration of Piled Raft Foundation for a High-Rise Building by using FEM

Pratish Kannaujiya[1] and Vijay Kumar Srivastava[2]

[1] Civil Engineering Department, Buddha Institute of Technology, Gorakhpur 273209, India

[2] Civil Engineering Department, Buddha Institute of Technology, Gorakhpur, 273209, India

Abstract. In this paper, the numerical simulation is performed on two piled raft configurations having uniform loading on a piled raft foundation. To investigatethe interaction between various parameters of pile soil foundation with varying components of its length of pile, diameter and raft thickness. These componentsare critical for increasing the foundation's bearing capacity. It's important aspectto achieve a reliable design to optimize piled raft foundations subjected to uniformed load- settlement behaviors of the curves. These load – settlements are basically depending on the changing of their components piled raft foundation. These component of piled raft foundations to choose an optimal selection of embedded length of pile (L/dP), normalized diameter of pile (L/dP) and normalized raft thickness (tR/dP). The effect of load-settlement on distributions of shear forceand bending moments is also investigated. In this study, a finite elements methods based on numerical tools ELPLA software is used for numerical simulation. The study's findings validate the numerical analysis for the calculation of proper pile configuration arrangement, as a result of settlements. It will be generated asformation of contours patterns as shear stress and bending moments on the rafts,may be minimized with the total pile length. The conclusion drawn from the studyvalidates the numerical analysis for the computation of proper pile arrangement can results in total and differential settlements, as well as generated shear stress and bending moments on the rafts, are all being reduced, with identical total pile length. Moreover, it captures the contours patterns of different behaviors of piledraft settlements, maximum shear force and maximum moments.

1. Introduction

Generally, high-rise buildings are frequently constructed on piled raft foundation whichis needed to withstand with vertical, lateral as well as overturning loadings. High rise buildings should also be designed for wind and seismic loadings. The foundation carries a large amount of axial stress on the earth in the case of such structures. So, it is important in the case of high-rise buildings that the foundation should be designed such that It can withstand superstructure loads while also transferring them to the earth. Moreover, it can be effectively supported the stability of the buildings. In some re- searchers to design of conventional and traditional methods that are used to design to resist the heavy loads on pile foundations [1, 2, 3]. It is actually a hybrid of a pile and raft foundation, consisting of three components: raft, piles, and subsoil. This type of foundation provides sufficient bearing capacity to the soil can also control settlement. The use of effective methods for designing pile raft foundations for a high-rise structurein order to reduce settlements and increase the foundations' bearing capacity. The weight is divided between piles and raft in a PRF, so raft and piles take on some of theload from the superstructure [4]. The load settlement curve of the piled raft has been computed based on the stiffness of the structures, and the design of the piled raft foundation is generally detailed in several phases [4, 5]. The analysis of the piled raft foundation for total and differential settlement, hybrid method is to be used [6] and analysis based on the group piling as well as optimum design of their rafts are analysed by manyresearchers in [7, 8]. The numerical analysis is done by a finite element method in which raft is modelled as a plate and piles treated as non-linear springs [9]. The interaction between the piled raft with deep foundations of various parameters are varied in terms of lengths parameters are subjected to their various loadings conditions by usingthe finite layer method [10].

Piled raft foundations (PRF) are in short supply has been highlighted in earlierresearch to enhance the overall stability of the building. Furthermore, numerous criticalelements that affect foundation stability, Factors including raft thickness, pile length, and diameter affect stability. These elements influence the overall stability of structuralsystems; however, the aforementioned parameters must be chosen carefully. It also leads to the construction of a cost-effective piled raft foundation. For a 20-story super-structure with a total height of 60 m, detailed design for pile raft foundation systems with the best pile raft design by pile arrangement scheme is required. In this story, the total load of building is 330 MN is calculated manually for using various types Indian codes [11, 12, 13, 14]. So, in this paper the numerical studies are based on the FEM. This was done to see how the length of the pile, the diameters of the piles, the thicknessof the raft, the number of piles, and their configurations affected the piled raft foundation.

2. Modelling and analysis

The overall designing is based on piled raft foundation by using traditional methods, in case of the applied load was carried out either by the pile or by the raft, consideringall the factors of safety in each of the following cases. The analysis of piled raft performance the proposed model is evaluated and three broad classes of several different methods are computer-based have been developed by [15, 16, 17, 18]. This parametricstudy was performed using the finite element software ELPLA. These tools are used to analysed the parameters of variety of sub model to evaluate the performance of a piledraft foundation. The numerical analysis is based on the different types of subsoil models that are specially designed for 3D continuum methods. It can be constructed to be analogous to an elastic foundation is analysed by finite element tool ELPLA. The piled raftfoundations were performed to validate the models with dimensions, soil properties, and loads that were carried out by [19] is shown in Fig. 1. The computed total load settlement (total load of 18 MN) relationships considered in this analysis are compared.The overall studies are performed to ELPLA software. The various methods for the center of the raft on 9 identical piles, by each column. Its components of high-rise building frame as seismic loading conditions are analysed with their case studies of different area [20].

In this present study, the raft was modelled as a concrete element with an area of the piled raft is 20 m × 20 m with loading applied on their rafts. It investigates the maximum settlement, bending moment on a pile raft foundation. In this study to investigate the two cases of piled raft configurations that are taken to analyse the effective foundation. The optimal selection of above parameters for designing leads to an economical design of pile raft foundation. The numerical simulations are based on the piled raft foundations in ELPLA software was carried out by [21]. In piled raft foundations are basically design with their raft as well as piled with varying all parameters and goodaggregate between their results of settlements [22]. In this present study to investigatesthe changes the various parameters such as the number of piles in a different pile con- figuration, pile length (L_P), pile diameters (d_P), and raft thickness (t_R). These parametersare mostly based on the design for the foundation systems was carried out by using twodifferent components such as raft and piles are isolated form of embedded in the soil layered on the foundation systems. The conditions of the soils and foundations proper-ties are presented in the soil layers are available for pilling in case of loose, medium and dense soil are 0-4 m, 4-15 m, 15-35 m. The details Model configuration and properties are used in this analysis as Fig. 1 and all the various materials are used for numerical analysis of the pile raft foundation as shown in Table 1. So, a compressive numerical study based on finite element tool ELPLA software. It was carried out to seeif there was any influence on the length of the piles in a pile group, effects on diametersof piles on a pile group, the effect on spacing between the piles, and the effect on raft thickness.

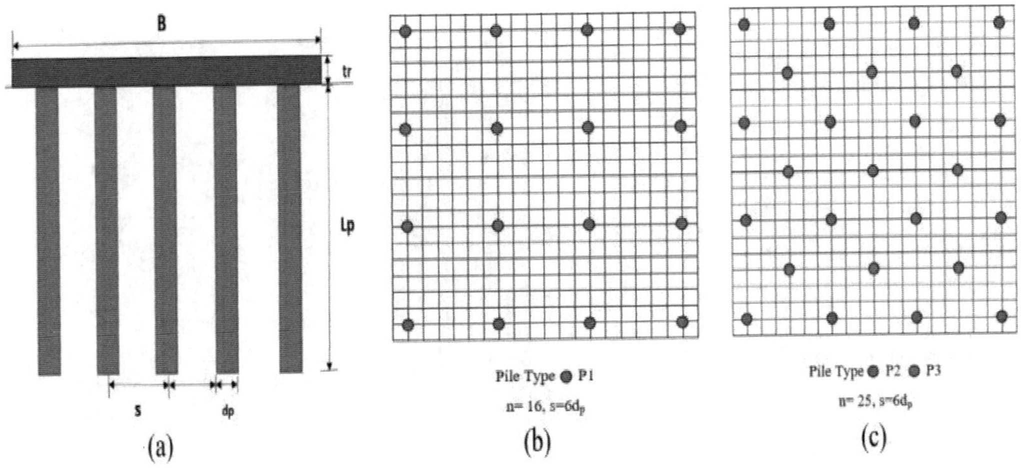

Figure 1. (a) model conditions with piles and rafts, (b) configuration of pile A, and (c) configurationof pile B

Table 1. Properties of soil types, Pile and raft

Parameter	**Soil type**			**Pile**	**Raft**
	Loose	**Medium dense**	**Dense**		
Youngs modulus, E(MPa)	10	20	24	3×10^4	2×10^4
Unit weight (kN/m^3)	16	18	20	25	25
Friction angle, ϕ (°)	24	32	37	-	-
Cohesion, c (kN/m^3)	1	2	2	-	-
Poisson's ratio, μ	0.3	0.3	0.3	0.18	0.25

In this study to investigate the two different types of pile configurations A and B are analysed with appropriate parameters. The configurations of pile A and B with the total number of piles in these configurations are 16 and 25 piles are analysed. In configurations (B) with 9 piles are types (P2) and (P3) with constant pile length 10m. It also changes the various parameters of pile configurations with different diameter ofpiles and raft thickness to analyse the behavior of maximum and differential settlements. It changes the parameters of pile raft that have been greatly affected on the maximum settlement are reduced and the bearing capacity of the foundation has been im- proved. Furthermore, to investigate the optimal selection of the leading parameters to an economical design of the pile raft foundation. In detail model of pile configurations (A) and (B) are shown in Fig. 2 and Fig. 3.

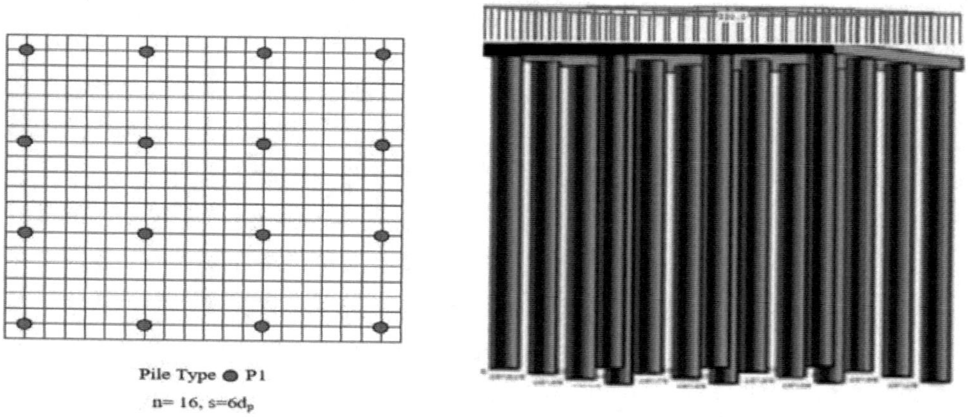

Figure 2. Pile configurations (A) with magnified view of piled raft foundation with loadings

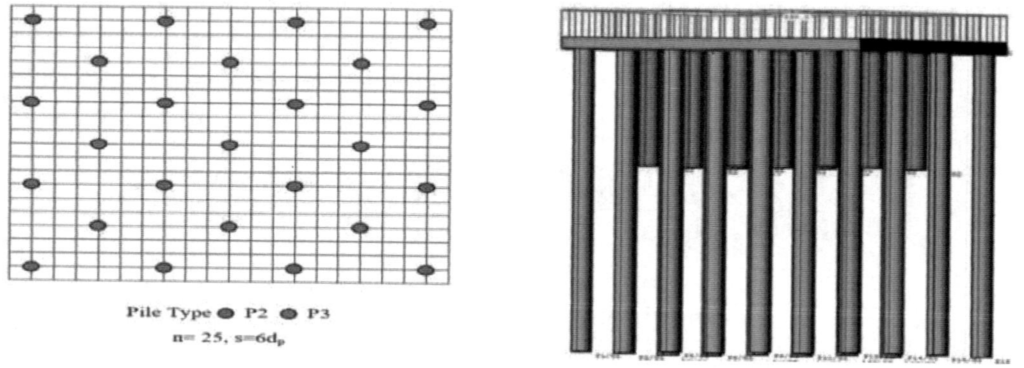

Figure 3. Pile configurations (B) with magnified view of piled raft foundation with loading

So, the above parameters are based on numerical studies that used the finite element approach to evaluate the influence of pile embedded length., effect of diameter ofpiles, and effect of raft thickness to an economical design of pile raft foundations. In the first case study to variation of pile type (P1) as a normalized pile length (Lp/dp) withconstant diameter of pile 1.0 m from 10 to 30. In the second case study to analyse the behavior of pile raft is subjected to the variation of pile type (P2) lengths are varies from 10 to 30m and constant the pile length (L = 10m) in pile type (P3). It also constantthe diameter of the piles and raft thickness is 1.0 m. The variation of embedded lengthof piles (L_1/L_2) ranging from 1 to 3 in configuration B. These variations of pile lengthsare analysing to choose an optimum selection of embedded pile length. Furthermore, to investigate the diameter of piles to an optimum selection of diameters is required to satisfy the both design criteria of the normalized diameter of piles and normalized pilespacing. The normalized diameter of pile (Lp/dp) varies in configuration A and B from 16 to 52 and 1.5 to 5.5. It also varying the normalized raft thickness (t_R/dP) in configuration A and B from 0.5 to 1.5 and 0.4 to 1.2. These variation of increasing the raft thickness is effectively reducing the differential settlements and also increasing maxi- mum moment. It is noted that increasing the raft thickness to resisting the punching shear from their loading conditions.

3. Results and analysis

In this study, the numerical analysis is conducted to evaluate the optimum designof pile configuration A and B for the designing of various parametric studies is done. In configuration A, the variation of normalized pile length from 10 to 30 with constantdp as a 1.0 m. These studies

were also investigated the behavior of pile raft was analysed in pile configuration B with varying the length (L_1) ranging from 10 m to 30 m in a piletype (P2). It also constant the pile length (L_2) is 10 m in a pile type (P3) and constant the pile diameter and raft thickness are 1m considered in this present study. In the lengthof pile required to improve the bearing capacity of the foundation by varying the pile length from 10 m to 30 m i.e., the variation of embedded pile length (L_1/L_2) ranging from 1 to 3 were investigated in these configurations. The provisions of the optimum selection of the length of pile (L_1) are investigated for reducing the maximum and differential settlement for the stability of the structures. It shows that the variation of settlements in pile configuration A and B as shown in Fig. 4.

In figures, are clearly shows that the settlement decreases with increases in pile length. The provisions of the proportion load borne by the piles are unaffected by the pile length embedded in the ground, varying the length of the pile with no major difference between the maximum length, while as a settlement. In the configuration A, the maximum settlement curve decreasing throughout their variation of normalized embedded pile length (L/dp) 10 to 30. Its aforementioned study to selection of optimum maximum settlement is achieved at 26. In this case, to consider here, there is little or no benefit in increasing the embedded pile length (L_1/L_2) above about 2.6 in pile configuration B.

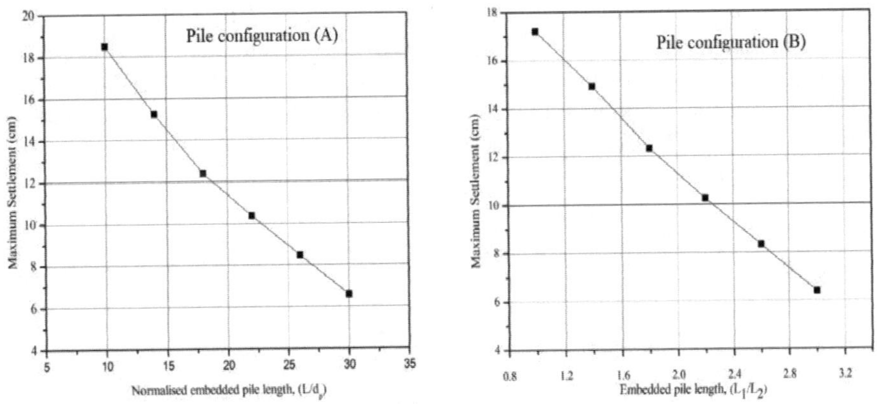

Figure 4. maximum settlement with the length of piles in pile configuration A and B

Further to investigate the diameter of piles for selection of optimum diameter of piles to improve the overall stability of the foundations and in a previous case study toobtain the optimum pile length (i.e., L/dp = 26 and L_1/L_2 = L= 2.6). It considered that the embedded pile length at (L= 2.6) and constant the values of raft thickness as 1 m inthis analysis further investigate the diameter of pile. In this analysis, the interaction factor for the vertically loaded pile is 332 kN/m^2 applied on their piled raft. The behavior of loading conditions in the raft and raft load transfer through their piles. The maximum settlement curves are found in forms of the normalized diameter of pile (L/dp) ranging from 16 to 52 and 1.5 to 5.5 in configuration A and B. The settlement of the curve is showing the relationship between the maximum settlement and the piles' normalized diameter as shown in Fig. 5. The variation of the differential settlement is influenced by changing the diameter of pile is greatly influences the performance of the pile raft foundations as shown in Fig. 9. In configuration A and B to achieved that the optimum design of pile diameters is adopted as 1.0 m (i.e., /dp = 21 and L/dp = 2.1) because the diameter of the piles increases but according to the applied load there is also reduced the settlements beyond which very little additional amount settlement would be obtained.

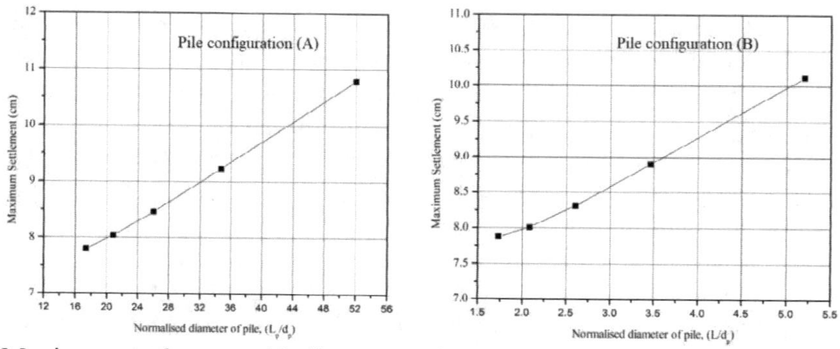

Figure 5. Maximum settlement with diameters of pile (Lp/dp) in pile configuration (A) and (B)

Also, further estimation of the pile raft foundation, the parameters are usually affected on the overall stability of the foundation systems. In the case of pile configuration, the raft thickness is a very important thing to design a pile raft system and pro- visions of the optimum raft thickness. To investigate the raft thickness on behalf of thedesign of piled raft configurations to estimate the variation of the raft thickness rangingfrom 0.5-1.5 m i.e., raft thickness (tR/dp) ranging from 0.4 to 1.2. The maximum settlement with the raft thickness is assessed for tR/dp ranging from 0.4 to 1.2 as shown in Fig. 6. In a pile raft configuration for providing a thin raft, the raft thickness has littleeffect on the maximum settling. The result from the linear analysis was obtained by thevariation of (tR/dp) to achieve an optimum value of the raft thickness is 1.0 m. In ac- cording to their variation of normalized raft thickness is obtained on the behalf contourspatterns of a normalized raft thickness corresponding to their maximum settlement as considerable at 1.0 m as a raft thickness shown in Fig. 7 and Fig. 8.

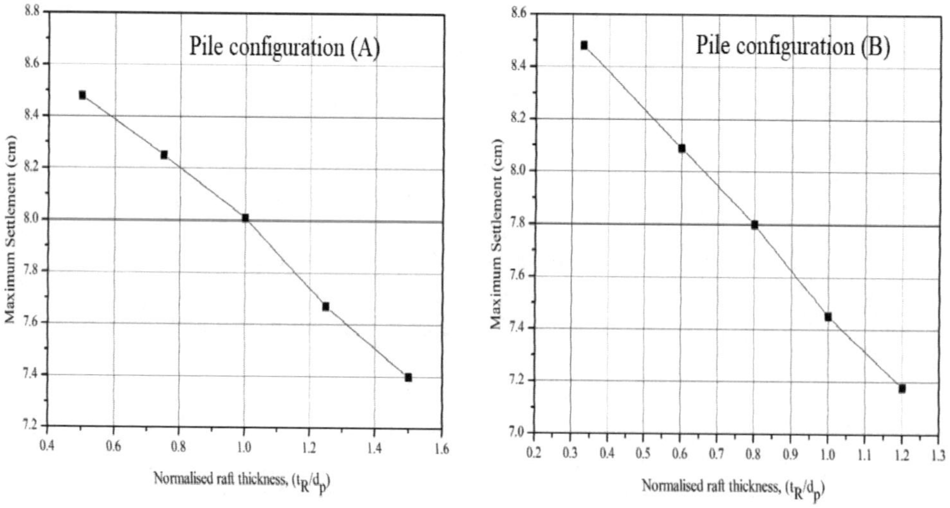

Figure 6. The curve is drawn between the Settlements v/s raft thickness (t_R/dp) in configurationof pile (A) and (B)

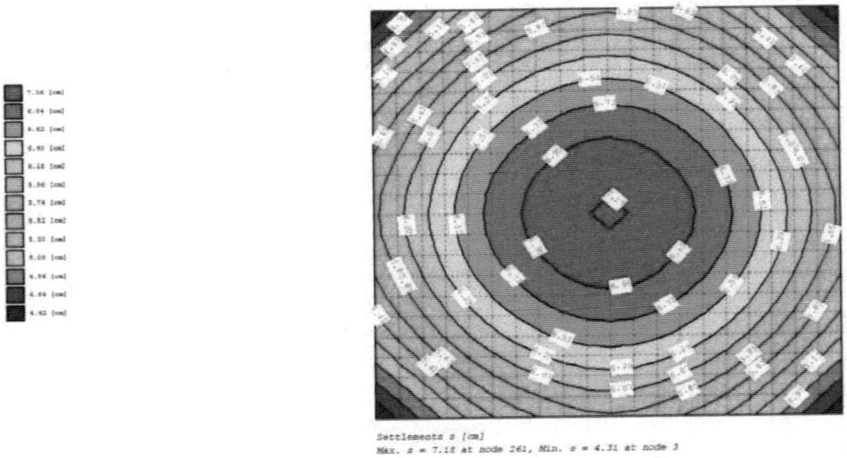

Figure 7. Contour pattern of normalised raft thickness with the optimum maximum settlement inpile configuration (A)

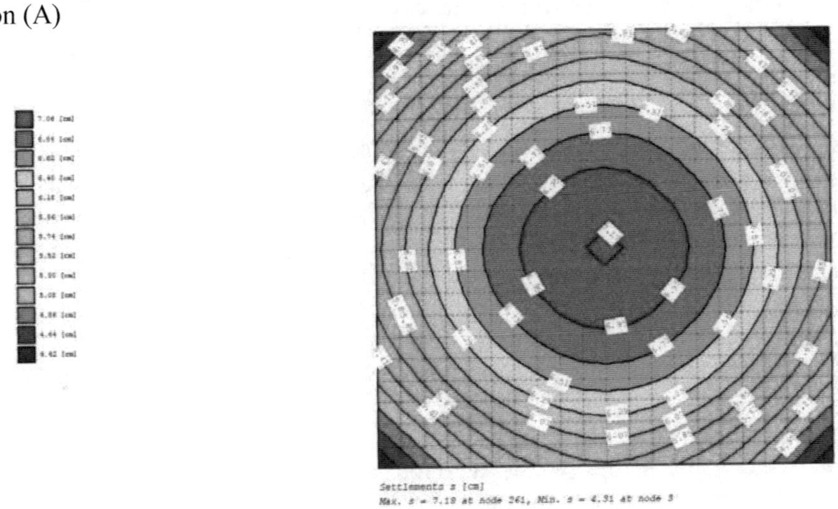

Figure 8. Contour pattern of normalised raft thickness with the optimum maximum settlement inpile configuration (B)

In the case of the pile configuration, raft thickness increases significantly with de- creases in the settlement. The provisions of each load are transferred by their piles are insensitive to the raft thickness and in a case, to consider the little amount or no benefitin increasing the thickness of rafts above about 1.0 m. It also shows that the variation of maximum moments curve also increases with increases the thickness of rafts is shown in Fig. 9. It shows that the selection of contour patterns at ($t_R/dp=1$) an optimummaximum moment in configuration of pile (A) and (B).

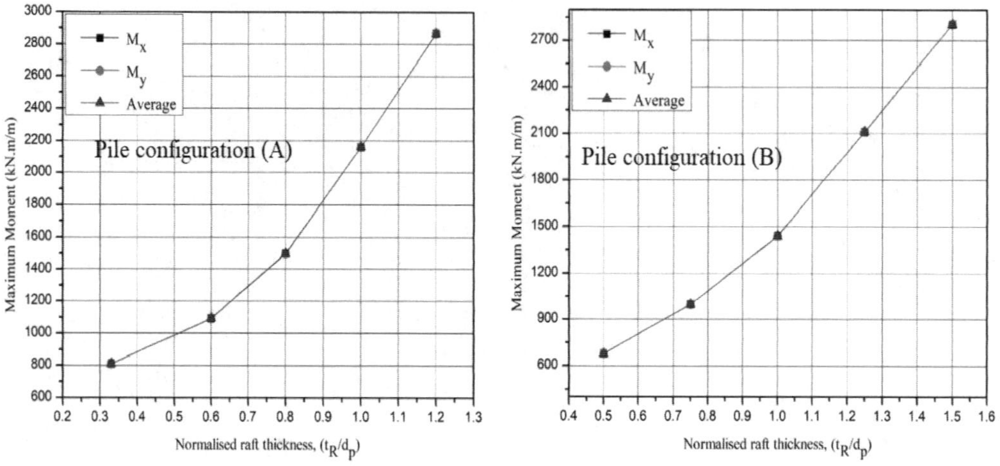

Figure 9. maximum moment increases with increases of thickness of rafts (t_R/dp) in pile configuration (A) and (B)

4. Conclusion

This paper shows that the numerical modelling of various parametric studies is done. Itemploys pile raft foundations to evaluate the total efficiency in managing settlement and enhancing the foundation's bearing capacity while developing high-rise buildings.To analyse the different types of parameters for designing optimum piled raft configurations systems. This study shows that optimizing the parameters and configuration provide the design aspect for pile raft with optimum consumption of the pile materials are as follows.

1. It is most economical design to improves a piled raft foundations to observedthat piles length (L/dp = 26 & L_1/L_2 = 2.6) has a great influence on improves on a foundations designs of the capacity are improved.
2. In a comparison of configuration of pile, A is more effective than A, it reduces(34%) the total settlement of foundations. In the piled raft foundation to use different configurations of piles to decreases the settlements and also improves the bearing capacity of the foundations.
3. It also found that improving the foundation of structures should be an optimumdiameter of the piles should be taken as (dp = 1.0 m). It also would be taken asthe normalized diameter of pile was achieved at (Lp/dp = 2.1). It reduces the settlement in pile raft configuration A is 7 % and pile raft configuration B is 27%. It provides the pile raft configuration B, i.e, it a economical other than configuration of pile A.
4. It is noted that the optimum thickness of raft (t_R) has to decrease the total settlement of the buildings and provided an optimum normalized raft thickness (t_R/dp = 1.0) in both configuration of piles A and B.
5. There are recommended for using configuration of pile B as compared to bothconfiguration of pile A and B. It is slightly economical as compared to pile raft configuration A.

5. Reference

1. Das, Braja M.: Principles of foundation engineering. Seventh edition (1984).
2. Tomlison, M., Woodward, J.: Pile design and construction practice. Fourth edition https://doi.org/10.1201/b12838 (1994).
3. Bowles, J.E.: Foundation Analysis and Design. 5th Edition, McGraw-Hill Companies, Inc., Singapore (1996).
4. Poulos, H. G.: High-Rise Building Foundations-A Limit State Design Approach. Art of Foundation Engineering Practice. doi:10.1061/41093(372)25 (2010).
5. Poulos, H. G.: Piled raft foundations: design and applications. Géotechnique, 51(2), 95–113. doi:10.1680/geot.2001.51.2.95 (2001).
6. Horikoshi, K., Randolph, M.: Estimation of overall settlement of piled rafts. Soils and Foundations, 39(2), 59-68 (1999).
7. Randolph, M. F.: Design Methods For Pile Groups and Piled Rafts. Proc. 13th Int. Conf. SoilMech. Found. Engng, New Delhi 5, 61-82 (1994).
8. Horikoshi, K., Randolph, M.: A contribution to optimum design of piled rafts. Geotechnique: international journal of soil mechanics, 48(3), 301-317. https://doi.org/10.1680/geot.1998.48.3.301 (1998).
9. Russo, G.: Numerical analysis of piled rafts. International Journal for Numerical and Analytical Methods in Geomechanics, 22(6), 477–493. doi:10.1002/(sici)1096-9853(199806)22:6<477::aid-nag931>3.0.co;2-h (1998).
10. Chow, H. S. W., Small, J. C.: Behaviour of piled rafts with piles of different lengths and diameters under vertical loading. Advances in Deep Foundations. doi:10.1061/40778(157)20 (2005).
11. IS 1893 Part 1.: Criteria for earthquake resistant design of structures: General provisions and buildings. Bureau of Indian Standards, New Delhi (2016).
12. IS 875 Part 1.: Code of practice for design loads (other than earthquake) for buildings and structures-dead loads-unit weights of building materials and stored materials. Bureau of Indian Standards, New Delhi, India (1987).
13. IS 875 Part 2.: Code of practice for design loads (other than earthquake) for buildings and structures-imposed loads. Bureau of Indian Standards, New Delhi, India (1987).
14. IS 875 part 3.: Design loads (other than earthquake) for buildings and structures-code of practice for wind loads. Bureau of Indian Standard, New Delhi, India (2015).
15. H.G Poulos *et.al.*: Comparison of some methods for analysis of piled raft. Proc.14 ICSMFE, Hamburg, 2:1119-1124 (1997).
16. Poulos, H. G., Davis, E. H.: Pile foundations analysis and design. John wiley and sons, p. 397 (1980).
17. Butterfield, R., Banerjee, P. K.: The Elastic Analysis of Compressible Piles and Pile Groups. Géotechnique, 21(1), 43–60. doi:10.1680/geot.1971.21.1.43 (1971).
18. Burland, J. B.: Piles as Settlement Reducers. Keynote address, 18Th Italian congress on soil mechanics, Pavia, Italy (1995).
19. Rabiei, M.: Effect of Pile Configuration and Load Type on Piled Raft Foundations Performance. Deep Foundations and Geotechnical In Situ Testing. doi:10.1061/41106(379)3 (2010).
20. Poulos, H. G.: Lessons learned from designing High-Rise Building Foundation. 19th SEAGC& 2 AGSSEA (2016).
21. El Gendy M, El Gendy A.: Analysis and design of raft and piled Raft-Program ELPLA. GEOTEC Software Inc., Calgary (2020).
22. Kannaujiya, P., Jaiswal S., Chauhan V.B.: Studies on the Piled Raft Foundation for a High-Rise Building Using Finite Element Modeling. Advance in Geo-Science Geo-Structure. Lecture Notes in Civil Engineering, Vol 154. Springer, Singapore. https://doi.org/10.1007/978-16-1993-926 (2021).

IOP Conf. Series: Materials Science and Engineering 1236 (2022) 012007 doi:10.1088/1757-899X/1236/1/012007

Design of Pile Foundation for Distillation Column

Rajkumar R. Betkar[1] and Atteshamuddin S. Sayyad[2][0000-0002-3702-4167]
[1]M.Tech. Student, Department of Structural Engineering, Sanjivani College of Engineering, Kopargaon, Savitribai Phule Pune University, Kopargaon-423603, Maharashtra, India
[2]Professor, Department of Structural Engineering, Sanjivani College of Engineering, Kopargaon, Savitribai Phule Pune University, Kopargaon-423603, Maharashtra, India
rajkumarbetkar92@gmail.com

Abstract. Equipment foundation is one of the most critical facility in every petrochemical, fertilizer, oil and gas plants. This equipment comprises distillation columns, vertical vessels, horizontal vessels, high pressure exchangers, reactors etc. Out of this equipment, distillation column is one of the most critical equipment due to its large diameter and height hence resulting into heavy dead load and heavy transverse loads at the base of column. Hence proper foundation system should be designed to cater all these loading without affecting the strength and serviceability of the foundation system. In most of the cases due to poor bearing capacity of soil one need to go for pile foundation. This paper summarizes the Indian code and process industry practices, design loads and load combinations to be adopted for design of pile foundation. One sample example is solved to evaluate the no of piles for column foundation by assuming suitable loading and vessel data.

1 Introduction

Equipment foundation is one of the key components in every refinery. This equipment consists of static as well as dynamic/rotating equipment. Static equipment consists of horizontal vessel, vertical vessels, process column, stacks, static pumps, reactors, heater, tanks, etc. whereas dynamic equipment consists of dynamic pumps, compressors, etc. Foundations for these equipment varies from single isolated octagon footing, combined open foundations, single pile cap or combined pile cap supporting one or more than one vessel. In this paper, author presented the design philosophy, design load cases and load combinations to be considered for design of industrial vertical equipment followed by one solved example to show the calculation for evaluation of pile numbers for large process column supporting on single isolated pile cap [1-5].

Fig.1 shows 3D view of distillation column.

2 Design Criteria

Design criteria for vertical vessel foundation is given in PIP STE03350.

3 **Design** *Loads*

In the design of column/vertical vessel foundation mainly dead load, live load, wind load and seismic loads are considered. The definition of these loads as per PIP STE03350 [1] is defined as below.

a) ***Dead Loads (DL)*:** Dead load consists of following,
   i. Structural dead load: This consists of weight of foundation and overburden soil weight including surcharge if any.
   ii. Equipment Erection load: which is only fabricated weight of equipment without any attachments to it and this weight is provided by the vendor in vendor drawing.
   iii. Equipment empty load: equipment empty dead load includes erection weight plus all the attachments, trays, internals, bundle, insulation, fireproofing, agitators, piping, ladders, platforms, etc.
   iv. Equipment operation dead load: it is the empty load of the equipment plus maximum weight of content during normal operation.
   v. Equipment test dead load: it is the weight of the vessel during test condition which includes empty weight of the vessel plus test medium contained in it.

b) ***Live Loads (LL)*:** Live loads are the loads which are generated by personnel, tools, miscellaneous equipment, movable partitions, wheel loads, etc. In case of vertical vessels, process columns the effects of live loads are less compared with other loads.

c) ***Wind Loads (WL)*:** Wind loads should be calculated in accordance with PIP STC01015 and IS 875 part 3. While calculating the wind load due consideration should be given to the factors such as exposed ladder, piping, spoiler, platforms, etc. on vessel. While designing the foundation the effect of diagonal wind loads should also be taken into account.

d) ***Earthquake Loads (EL)*:** Earthquake loads should be calculated in accordance with PIP STC01015 and IS 1893 part 1 & 4. Seismic base shear should also be verified and should not be less than the limits given in IS 1893. Proper comparison should be made between site spectra and codal spectra and stringent among above should be followed. Depending upon the category of equipment MCE or DBE loading should be calculated & considered in the analysis & design of foundation system.

3.1 Load Cases

Following load cases are considered for the analysis and design of vertical equipment foundation.

1) Dead load (self-weight of foundation + surcharge soil weight if any)
2) Equipment empty/erection load
3) Equipment operating load
4) Equipment test load
5) Wind load in X-direction
6) Wind load in Z-direction
7) Seismic load (DBE) in X-direction
8) Seismic load (DBE) in Z-direction
9) Seismic load (MCE) in X-direction
10) Seismic load (MCE) in Z-direction

3.2 Load Combinations

With the reference of PIP STC01015, Indian standard code IS456:2000, IS 800:2007& IS 1893 author mentioned below the possible load combinations to be used for the design of vertical column

IOP Conf. Series: Materials Science and Engineering 1236 (2022) 012007 doi:10.1088/1757-899X/1236/1/012007

foundation. While preparing these load combinations below mentioned scenarios are also taken into account [1-5].

- Effect of foundation uplift due to presence of shallower water table or minimum operating load with heavy lateral loads.
- Effect of both Design basis earthquake and maximum considered earthquake.

Table 1 and 2 shows load combinations to be used for serviceability and strength check respectively.

Table 1. Load combinations for serviceability checks

Sr. No.	DL	EE	EO	ET	WLX	WLZ	EE. EQX (DBE)	EE. EQZ (DBE)	EQX (DBE)	EQZ (DBE)	EQX (MCE)	EQZ (MCE)
1	1.0	1.0										
2	1.0	1.0			1.0							
3	1.0	1.0			-1.0							
4	1.0	1.0				1.0						
5	1.0	1.0				-1.0						
6	1.0	1.0					1.0					
7	1.0	1.0					-1.0					
8	1.0	1.0						1.0				
9	1.0	1.0						-1.0				
10	1.0		1.0									
11	1.0		1.0		1.0							
12	1.0		1.0		-1.0							
13	1.0		1.0			1.0						
14	1.0		1.0			-1.0						
15	1.0		1.0						1.0			
16	1.0		1.0						-1.0			
17	1.0		1.0							1.0		
18	1.0		1.0							-1.0		
19	1.0		1.0								1.0	
20	1.0		1.0								-1.0	
21	1.0		1.0									1.0
22	1.0		1.0									-1.0
23	1.0			1.0								
24	1.0			1.0	0.25							
25	1.0			1.0	-0.25							
26	1.0			1.0		0.25						
27												
28	1.0			1.0		-0.25						

Table 2. Load combinations for strength design

Sr. No.	DL	EE	EO	ET	WLX	WLZ	EE. EQX (DBE)	EE. EQZ (DBE)	EQX (DBE)	EQZ (DBE)	EQX (MCE)	EQZ (MCE)
1	1.5	1.5										
2	1.5	1.5			1.5							
3	1.5	1.5			-1.5							
4	1.5	1.5				1.5						
5	1.5	1.5				-1.5						
6	0.9	0.9			1.5							
7	0.9	0.9			-1.5							
8	0.9	0.9				1.5						
9	0.9	0.9				-1.5						
10	1.5	1.5					1.5					
11	1.5	1.5					-1.5					
12	1.5	1.5						1.5				
13	1.5	1.5						-1.5				
14	0.9	0.9					1.5					
15	0.9	0.9					-1.5					
16	0.9	0.9						1.5				
17	0.9	0.9						-1.5				
18	1.5		1.5									
19	1.5		1.5		1.5							
20	1.5		1.5		-1.5							
21	1.5		1.5			1.5						
22	1.5		1.5			-1.5						
23	1.5		1.5						1.5			
24	1.5		1.5						-1.5			
25	1.5		1.5							1.5		
26	1.5		1.5							-1.5		
27	1.5		1.5								1.5	
28	1.5		1.5								-1.5	
29	1.5		1.5									1.5
30	1.5		1.5									-1.5
31	0.9		0.9		1.5							
32	0.9		0.9		-1.5							
33	0.9		0.9			1.5						
34	0.9		0.9			-1.5						
35	0.9		0.9						1.5			
36	0.9		0.9						-1.5			
37	0.9		0.9							1.5		
38	0.9		0.9							-1.5		
39	0.9		0.9								1.5	
40	0.9		0.9								-1.5	
41	0.9		0.9									1.5
42	0.9		0.9									-1.5

4 Design Example

4.1 Loading data:

Table 3. shows loading at top of pedestal

Weight	Erection	108950	kgf
	Operation	242250	kgf
	Full water	253700	kgf
Horizontal force	By wind	15350	kgf
	By earthquake DBE	144050	kgf
	By earthquake MCE	174900	kgf
Overturning moment	By wind	127770	kgf-m
	By earthquake DBE	1582280	kgf-m
	By earthquake MCE	1921330	kgf-m

4.2 Vessel Data

Diameter = 3972 mm, Shell length= 11500 mm, Skirt height= 3100 mm Bolt circle diameter= 4374 mm, Bolt numbers and size = 60 Nos-M48, size Bolt length = 2470 mm, Bolt projection = 605 mm

4.3 Pile capacities data:

Table 4. shows pile capacity

Compression capacity	800	kN
Tension capacity	600	kN
Shear capacity	95	kN
Allowable Increase In Vertical Capacity	wind	25%
	seismic	50%
Allowable Increase In Uplift Capacity	wind	25%
	seismic	25%
Allowable Increase In Shear Capacity	wind	25%
	seismic	25%

4.4 Calculations

Step 1: Sizing of pedestal

Considering octagonal pedestal, diameter of pedestal = BCD+12D

$= 4374+12\times48$

=4950 mm (consider 5000mm is the size of pedestal)

Step 2: Sizing of pile cap and piling layout

Considering octagonal pile cap of size 12 m and thickness 1 m.

Assuming no of piles required are 20 Nos (Fig.2 shows piling layout)

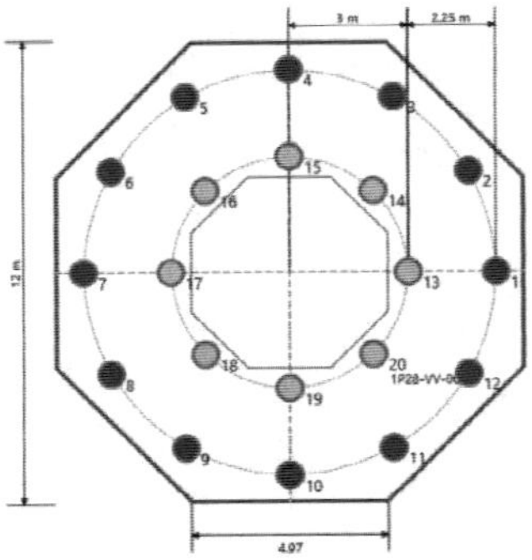

Fig.2 Loading calculations

In case of vertical equipment foundation, especially column foundation the governing load combination is DL+EO+ EQX (MCE)

Here dead load case includes

1. Weight of pile cap = 0.828×12×12×1×25= 2980.8 kN
2. Weight of pedestal = 0.828×5×5×2.3×25= 1190.25 kN

   (2.3 is ht. of pedestal from TOP of pile cap to TOP of grout/Bottom of base plate)
3. Weight of soil = 0.828 × (12^2-5^2) ×2×18= 3547.152 kN

   2 m is height of soil column from top of pile cap up to HPP Highest paving point level
4. Equipment operating weight = 2422.50 kN

   Total vertical load = 2980.8+1190.25+3547.152+2422.50 + push pull due to overturning moment.

Step 3: Push pull due to overturning moment

Overturning moment at base of pile cap due to earthquake MCE condition= 19213.30+1749× 3.3 = 24985 kN-m (3.3 is the height between top of pedestal to bottom of pile cap). Now push pull in the extreme pile due to this overturning moment will be,

$$P_v = \frac{\sum V}{N} + \frac{\sum V\, e_x\, c_x}{I_{yy}} \quad (1)$$

P_v= (10140/20) + (10140×2.463×5.25) / (201.451) = 507+650.866 = 1157.866 kN

Thus, total compressive force in an extreme pile is 1158 kN and tension/uplift force is 143.866 kN.

Step 4: Utility ratio check

a) Actual compressive force in extreme pile= 1158 kN
b) Allowable compressive force in pile due to earthquake condition = 800×1.5=1200 kN
c) Hence utility of pile in compression will be = 1158/1200= 0.965
d) Actual Tensile force in extreme pile= 143.866 kN
e) Allowable uplift force in pile due to earthquake condition = 600×1.25=750 kN
f) Hence utility of pile in Tension will be = 143/750= 0.190

g) Actual shear force in extreme pile= 1749/20=87.45

h) Allowable shear force in pile due to earthquake condition = 95×1.25=118.75 kN

i) Hence utility of pile in shear will be = 87.45/118.75= 0.736

5 Future scope of research

1) Analysis and design of pile cap using the truss analogy or using the bending theories.
2) Analysis and design of combined pile cap.
3) Effect of shear enhancement on design of pile cap.

6 Conclusions

Based on the design shown above, following conclusions are drawn.

- Maximum utility ratio of pile in compression is 0.965.
- Maximum utility ratio of pile in Tension is 0.190.
- Maximum utility ratio of pile in shear is 0.736.
- It is always preferable to restrict maximum utility ratio of pile up to 0.9 in any governing load combination.
- Industrial process columns mainly govern for DL+LL+EQ (MCE/DBE) depending upon the category of vessel.

7 References

1. PIP STE03350 Vertical Vessel Foundation Design Guide.
2. IS 456:2000 Plain And Reinforced Concrete- Code of Practice.
3. IS 800:2007 General Construction In Steel- Code Of Practice.
4. IS 1893 (PART 1):2002 Criteria For Earthquake Resistant Design Of Structures (Part 1 General Provisions And Buildings).
5. IS 1893 (PART 4):2005 Criteria For Earthquake Resistant Design Of Structures (Part 4 Industrial Structures Including Stack-Like Structures).

Bending of Functionally Graded Nanobeams using Hyperbolic Nonlocal Theory

A S Jape and A S Sayyad
Department of Structural Engineering, Sanjivani College of Engineering, Savitribai Phule Pune University, Kopargaon-423603, Maharashtra, India
japeanujast@sanjivani.org.in

Abstract. Hyperbolic nonlocal theory is applied in this paper to calculate deflections in functionally graded nanobeams under uniform load. The theory is developed using the work done principle in conjunction with Eringen's theory. Free stress conditions on the upper and lower surfaces are achieved by the current hypothesis. Deflection of beam is estimated using Navier's approach considering simple ends of the nanobeam. Solutions found in the literature are taken for the comparison purpose and found that the present findings are matching with the existing solutions.

Keywords: Bending analysis, hyperbolic nonlocal theory, Eringen's theory, functionally graded nanobeams.

1. Introduction

Nowadays functionally graded (FG) nanobeams are widely used in many industrial applications where those are subjected to bending actions due to transverse loads. Hence, it is important to accurately analyse the nanobeams under the bending action. The most widely used approach to model the nanobeams is Eringen's nonlocal theory [1]. Many researchers have developed nonlocal beam theories to analyse the isotropic nanobeams under bending action in conjunction with the Eringen's theory [2]. However, research on FG nanobeams is limited. In FG material, elastic constants are gradually varying along with the depth or length of the beam and hence it is treated an advanced composite material. Deflection of an isotropic beam utilizing classical and refined theories has been reported by Reddy [3]. Thai and Vo [4] have used sinusoidal function in the kinematics of nanobeams theory. Thai [5] has used third-order polynomial function for the modeling of nanobeams theory. Sayyad and Ghugal [6] have introduced inverse trigonometric nanobeams theory. Garg and Chalak [7] have developed novel higher order zigzag theory for analysis of laminated sandwich beams. Chalak, Zenkour, et al. [8] used zigzag theory for free vibration and modal stress analysis of FG-CNTRC beam under hygrothermal conditions. Belarbi, et al. [9] have used a novel shear deformation theory for nonlocal finite element model for the bending and buckling analysis of functionally graded nanobeams.

Garg et al. [10] studied hygro-thermo-mechanical based bending analysis of symmetric and unsymmetrical power-law, exponential and sigmoidal FG sandwich beams.

In this work, a hyperbolic nonlocal beam theory is applied to calculate central deflections in FG nanobeams under the action of uniform load on the top surface. The theory is formulated based on the work done principle and Eringen's model. Maximum deflections of nanobeams under uniform load are presented for isotropic material and functionally graded material. The present results are checked with previous research results and found closely matching with them for all nonlocal parameter values as well as the power-law index.

2. Materials and Methods

Geometry: Length of nanobeam is L, width is b and thickness is h (see Fig. 1)

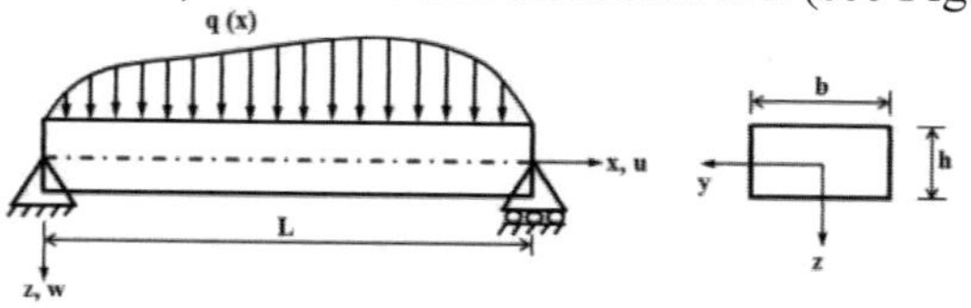

Fig. 1. FG nanobeams under consideration.

Material: The nanobeam is made up of functionally graded material. Elastic constants of the material are varied using the power-law of gradation stated in Eq. (1).

$$e(z) = e_m + (e_C - e_m)(0.5 + z/h)^p \tag{1}$$

where e is the elastic constant (E, μ, G); metal and ceramic materials are denoted by the letters m and c, respectively; and p is the power-law coefficient whose values range from 0 to ∞. The traction-free boundary conditions on the top and bottom surfaces of the beam are obtained using the hyperbolic shape function in the current nonlocal beam theory.

$$u(x,z) = u_0(x) - z\frac{\partial w_0}{\partial x} + \left[z\cosh\left(\frac{\xi}{2}\right) - \frac{h}{\xi}\sinh\left(\frac{z\xi}{h}\right)\right]\theta(x) \tag{3}$$
$$w(x) = w_0(x)$$

where, $\xi = 2.634$, u and w are the displacements in x- and z- directions; θ is the rotation. To account for the impacts of transverse shear deformation, hyperbolic shape function is employed in the axial displacement. The nonzero strains are obtained from the theory of elasticity.

$$\varepsilon = \varepsilon_1 + z\varepsilon_2 + \left[z\cosh\left(\frac{\xi}{2}\right) - \frac{h}{\xi}\sinh\frac{z\xi}{h}\right]\varepsilon_3$$
$$\gamma = \left[\cosh\left(\frac{\xi}{2}\right) + \cosh\left(\frac{z\xi}{h}\right)\right]\gamma_1 \tag{4}$$

where

$$\varepsilon_1 = \frac{\partial u}{\partial x},\quad \varepsilon_2 = -\frac{\partial^2 w}{\partial x^2},\quad \varepsilon_3 = \frac{\partial \theta}{\partial x},\quad \gamma_1 = \theta \tag{5}$$

The 1D Hooke's law is referred to calculate stresses within the domain of nanobeams.

$$\sigma = E(z)\varepsilon \quad \text{and} \quad \tau = \frac{E(z)}{2(1+\mu)}\gamma \tag{6}$$

where E(z) represents the FG beam's Young's modulus as it varies across thickness according to Eq. (1). To develop differential equations, the work done principle is combined with the present theory.

$$b\int_0^L \int_{-h/2}^{h/2} (\sigma\,\delta\varepsilon + \tau\,\delta\gamma)\,dz\,dx - \int_0^L q(x)\,\delta w\,dx = 0 \tag{7}$$

where *q(x)* represents the transverse load. The current theory's differential equations are generated by part wise integrating Eq. (7), collecting the multiplied differential terms of unknowns, and equating them with zero.

$$\begin{aligned} &A\frac{\partial^2 u_0}{\partial x^2} - B\frac{\partial^3 w_0}{\partial x^3} + C\frac{\partial^2 \theta}{\partial x^2} = 0 \\ &B\frac{\partial^3 u_0}{\partial x^3} - D\frac{\partial^4 w_0}{\partial x^4} + F\frac{\partial^3 \theta}{\partial x^3} = -q(x) \\ &C\frac{\partial^2 u_0}{\partial x^2} - F\frac{\partial^3 w_0}{\partial x^3} + H\frac{\partial^2 \theta}{\partial x^2} - J\theta = 0 \end{aligned} \tag{8}$$

where

$$(A,B,C,D,F,H,J) = b\int_{-h/2}^{h/2} E(z)\left[1, z, f(z), z^2, zf(z), f^2(z), (f'(z))^2\right]dz \tag{9}$$

The small-scale effect of nanobeams is explained using Eringen's theory. Nonlocal stresses may be calculated for functionally graded materials using the following relation.

$$\sigma - n\frac{\partial^2 \sigma}{\partial x^2} = E(z)\varepsilon \quad \text{and} \quad \tau - n\frac{\partial^2 \tau}{\partial x^2} = E(z)\gamma \tag{10}$$

Using Eq. (10) and Eq. (8); we can write final differential equations of nanobeams corresponding to the current theory.

$$\begin{aligned} &\delta u_0 : A\frac{\partial^2 u_0}{\partial x^2} - B\frac{\partial^3 w_0}{\partial x^3} + C\frac{\partial^2 \theta}{\partial x^2} = 0 \\ &\delta w_0 : B\frac{\partial^3 u_0}{\partial x^3} - D\frac{\partial^4 w_0}{\partial x^4} + F\frac{\partial^3 \theta}{\partial x^3} = -\left(q - n\frac{\partial^2 q}{\partial x^2}\right) \\ &\delta\varphi : C\frac{\partial^2 u_0}{\partial x^2} - F\frac{\partial^3 w_0}{\partial x^3} + H\frac{\partial^2 \theta}{\partial x^2} - J\theta = 0 \end{aligned} \tag{11}$$

Solution Method: Deflections in simply supported functionally graded nanobeam is estimated using the Navier approach which satisfies the simply supported end conditions exactly i.e. $N_x = w = M_x^b = M_x^s = 0$. The unknown variables of the current theory and the transverse load are assumed in the single Fourier series.

$$u_0=\sum_{m=1}^{\infty}u_m\cos rx, w_0=\sum_{m=1}^{\infty}w_m\sin rx, \theta=\sum_{m=1}^{\infty}\theta_m\cos rx,\ q(x)=\sum_{m=1}^{\infty}\frac{4q_0}{m\pi}\sin rx \quad (12)$$

where $r=m\pi/L$, (u_m,w_m,θ_m) are unknown coefficients, q_0 is the maximum intensity of uniform loading. Eq. (12) substituted into Eq. (11) to give solutions for the bending problem stated in Eq. (13).

$$\begin{bmatrix}-Ar^2 & Br^3 & -Cr^2\\ Br^3 & -Dr^4 & Fr^3\\ -Cr^2 & Fr^3 & -(Hr^2+J)\end{bmatrix}\begin{Bmatrix}u_m\\ w_m\\ \theta_m\end{Bmatrix}=-\frac{4q_0}{m\pi}\begin{Bmatrix}0\\ 1+nr^2\\ 0\end{Bmatrix} \quad (13)$$

3. Results and Discussions

The current theory is applied for the transverse central deflection estimation of uniformly loaded isotropic and FG nanobeams simply-supported at ends. Material properties used for the beam are E = 1 GPa and μ = 0.3 for isotropic material; and E_c = 380 GPa, E_m = 70 GPa, $\mu_c = \mu_m$ = 0.3 for FG material.

Table 1. Non-dimensional transverse deflection $\overline{w}=100EIw/(q_0L^4)$ of simple end isotropic nanobeams (*L=1nm, b=1nm, h=1nm, m=100*)

Theory	Nonlocal parameter (n)				
	0	1	2	3	4
Present	1.3351	1.4628	1.5904	1.7181	1.8457
Sayyad and Ghugal[6]	1.3394	1.4675	1.5956	1.7237	1.8517
Thai [5]	1.3346	1.4622	1.5898	1.7173	1.8449
Thai and Vo [4]	1.3345	1.4621	1.5897	1.7173	1.8449
Timoshenko [6]	1.3124	1.4382	1.5641	1.6901	1.8157
Bernoulli-Euler [6]	1.3020	1.4270	1.5522	1.6772	1.8022

Summary of transverse deflection in isotropic nanobeams for various values of the nonlocal parameters is present in Table 1. The present results are compared with Sayyad and Ghugal [6], Thai [5], Thai and Vo [4], Timoshenko, [6] and Bernoulli-Euler [6] This table illustrates that the current theory is in good accord with previous research results. The values of central transverse deflection for different values of power-law coefficients and nonlocal parameters are presented in Table 2. Theories of Sayyad and Ghugal [6] and Reddy [10] are used for the comparison. The present theory predicts the central deflection of FG nanobeams under uniform load with good accuracy.

4. Conclusions

This study highlights the use of hyperbolic theory function with arbitrary variables to predict the maximum central deflection of simply-supported nanobeams made up of homogeneous and functionally graded materials. It is concluded that the present theory is in good accord with previous research results while predicting the central deflection of uniformly distributed nanobeams.

Table 2. Non-dimensional transverse deflection $\bar{w} = 100E_m h^3 w / \left(q_0 L^4\right)$ of simple end FG nanobeams under uniform load (*L=10nm, h=1nm, m=100*)

Theory	n	Power law coefficients (p)			
		0	1	5	10
Present	0	2.9501	5.8958	9.0186	9.9391
	1	3.2321	6.4599	9.8801	10.888
	2	3.5141	7.0239	10.741	11.837
	3	3.7962	7.5880	11.603	12.786
	4	4.0783	8.1520	12.464	13.734
Sayyad and Ghugal [6]	0	2.9556	6.0409	9.2826	10.229
	1	3.2345	6.6093	10.154	11.189
	2	3.5134	7.1776	11.025	12.148
	3	3.7922	7.7462	11.897	13.108
	4	4.0711	8.3145	12.769	14.068
Reddy [10]	0	2.9501	5.8959	9.0204	9.9403
	1	3.2322	6.4599	9.8820	10.889
	2	3.5142	7.0240	10.743	11.838
	3	3.7963	7.5880	10.743	12.787
	4	4.0783	8.1521	12.466	13.736

5. Conflict of Interest

The authors state that the publication of this paper does not include any conflicts of interest.

References

1. Eringen, A.C.: On differential equations of nonlocal elasticity and solutions of screw dislocation and surface waves. Journal of Applied Physics 54, 4703-4710 (1983).
2. Eltaher, M.A., Khater, M.E., Emam, S.A.: Review on nonlocal elastic models for bending, buckling, vibrations, and wave propagation of nanoscale beams. Applied Mathematical Modelling 40, 4109–4128 (2016).
3. Reddy, J. N.: Nonlocal theories for bending, buckling and vibration of beams. International Journal of Engineering Science 45, 288-307 (2007).
4. Thai, H., Vo, T.P.: A nonlocal sinusoidal shear deformation beam theory with application to bending, buckling, and vibration of nanobeams. International Journal of Engineering Science 54, 58–66 (2012).
5. Thai, H.: A nonlocal beam theory for bending, buckling, and vibration of nanobeams. International Journal of Engineering Science 52, 56–64 (2012).
6. Sayyad, A.S., Ghugal M.Y.: Bending, buckling, and vibration analysis of

functionally graded nanobeams using an inverse trigonometric beam theory. International Journal of Nano Dimensions 12(2), 164-174 (2021).
7. Garg, A., Chalak, H.D.: Novel higher-order zigzag theory for analysis of laminated sandwich beams. Proceedings of the Institution of Mechanical Engineers, Part L: Journal of Materials: Design and Applications 235(1), 176-194 (2020).
8. Chalak, H.D., Zenkour, A.M., Garg, A.: Free vibration and modal stress analysis of FG-CNTRC beams under hygrothermal conditions using zigzag theory. Mechanics Based Design of Structures and Machines, An International Journal (2021)
9. Belarbi, M.O., Houari, M.S.A., Daikh, A.A., Garg, A., Merzouki, T., Chalak, H.D., Hirane. H.: Nonlocal finite element model for the bending and buckling analysis of functionally graded nanobeams using a novel shear deformation theory. Composite Structures 264, 113712 (2021)
10. Garg, A., Chalak, H.D., Belarbi, M.O., Zenkour, A.M.: Hygro-thermo-mechanical based bending analysis of symmetric and unsymmetric power-law, exponential and sigmoidal FG sandwich beams. Mechanics of Advanced Materials and Structures (2021)

Sustainable Development in Civil and Electrical Engineering (SDCEE 2021) IOP Publishing
IOP Conf. Series: Materials Science and Engineering 1236 (2022) 012009 doi:10.1088/1757-899X/1236/1/012009

Effect of Thickness Stretching on the Natural Frequencies of Layered Composite Beams

Nitin S. Naik[1[0000-002-5302-5581]] **and Atteshamuddin S. Sayyad**[2[0000-0002-3702-4167]]

[1] Department of Structural Engineering Sanjivani College of Engineering, Kopargaon (M.S.) India, Savitribai Phule Pune University, Pune.
[2] Department of Structural Engineering Sanjivani College of Engineering, Kopargaon (M.S.) India, Savitribai Phule Pune University, Pune.
nitin_adi59@rediffmail.com

Abstract. In this study, a new fifth order shear deformation theory is developed considering the effect of thickness stretching for the free vibration analysis of laminated composite beam. To account the effect of shear in transverse direction and normal de-formations, displacement field uses polynomial shape function in terms of thick-ness co-ordinate expanded up to order five. The virtual work equation is used to have the equations in terms of the unknowns and corresponding boundary conditions. The analytical solution for a laminated composite beam with simple sup-ports is developed using Navier's solution. To assess the validity of the theory developed, dimension less natural frequencies for laminated composite beam determined by the present model are compared with previously published results.

1. Introduction

Structures made of composite materials are very important and popular structures which are used in the modern engineering and aerospace industry because of their light weight and high strength to stiffness ratio. Such light weight structures are used extensively in civil, mechanical and transportation engineering. As these structures are made up of layers, they are subjected to number of challenging issues such as fiber breaking and separation of layers hence the correct prediction of stress variation through the thickness, defections, buckling loads, natural frequencies in case of member subjected to vibrations is required while analyzing such structures. Many researchers are working presently in the area and in the past many researchers have given analytical and solutions based on finite element method for the problems.

The well-established theory for thin beams was developed by Euler-Bernoulli [1-2], the theory neglects the transverse shear and stretching of thickness. To overcome the drawbacks of the Euler-Bernoulli beam theory, Timoshenko [3-4] developed a theory for beams considering the effect of shear deformation in transverse direction. This theory requires a shear correction coefficient due to assumption of a constant shear strain through the thickness of the beam. Reddy [5] has presented a higher-order theory, which is useful for the solution of different beam and plate problems. Sayyad and Ghugal [6] studied the thickness stretching effect on the natural frequencies of layered and sandwich beams with some results generated using different theories. Garg and Chalak [7] presented the bending and free vibration analysis of laminated beams under the static as well as free vibration conditions using a novel higher-order zigzag theory. Chalak et al. [8] performed bending and vibration analysis of FG softcore sandwich beams using finite element based higher-order zigzag theory. Chalak et al. [9] studied the free vibration and modal stress analysis of FG-CNTRC beams subjected to hygrothermal conditions using C-0 finite element based zigzag theory. In this article a fifth-order shear and normal deformation theory (FOSNDT) is used for the vibration analysis of layered composite beam which was developed by Naik and Sayyad [10-12] for the study of layered plates carrying mechanical and hygro-thermal environmental load.

2. Methodology

The beam under consideration is having a rectangular cross section with width '*d*', depth '*h*' and length '*L*' (See fig. 1).

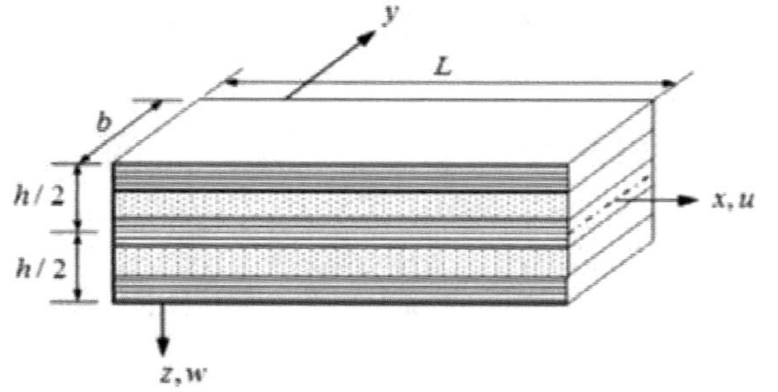

Fig.1 Geometry of the beam.

For the present theory a polynomial displacement field is assumed as stated in Eq. (1)

$$u = u_0 - z\left(\partial w_0/\partial x\right) + (z - 4z^3/3h^2)\phi_x + (z - 16z^5/5h^4)\psi_x$$
$$w = w_0 + (1 - 4z^2/h^2)\phi_z + (1 - 16z^4/h^4)\psi_z \tag{1}$$

where u and w are the displacements along x and in the direction of thickness respectively. ϕ_x and ψ_x are the shear slopes. ϕ_z and ψ_z are the slopes which account for transverse normal deformations. The strains are obtained using the basic equations of theory of elasticity. Eq. (2) gives the strain components.

$$\varepsilon_x = \frac{\partial u}{\partial x},\ \varepsilon_z = \frac{\partial w}{\partial z},\ \text{and}\ \gamma_{xz} = \frac{\partial u}{\partial z} + \frac{\partial w}{\partial x} \tag{2}$$

For the present study a 2-D Stress-Strain relationship is used to determine the stresses. The constitutive relations for the kth layer are stated as

$$\begin{Bmatrix} \sigma_x \\ \sigma_z \\ \tau_{xz} \end{Bmatrix}^{(k)} = \begin{bmatrix} \bar{Q}_{11} & \bar{Q}_{13} & 0 \\ \bar{Q}_{13} & \bar{Q}_{33} & 0 \\ 0 & 0 & \bar{Q}_{55} \end{bmatrix}^{(k)} \begin{Bmatrix} \varepsilon_x \\ \varepsilon_z \\ \gamma_{xz} \end{Bmatrix}^{(k)} \tag{3}$$

where $\bar{Q}_{ij}$ are the elastic constants in the *z-z* plane. σ_x is the stress acting in the *x*-direction, σ_z is the stress in the direction of thickness i.e. z and τ_{xz} is the shear stress acting along the direction of thickness. The reduced elastic constants in Eq. (4) are given below.

$$\bar{Q}_{11} = \frac{E_1}{1 - \mu_{13}\mu_{31}},\ \bar{Q}_{13} = \frac{\mu_{13}E_3}{1 - \mu_{13}\mu_{31}},\ \bar{Q}_{33} = \frac{E_3}{1 - \mu_{13}\mu_{31}},\ \bar{Q}_{55} = G_{13} \tag{4}$$

The Principle of virtual work as stated in Eq. (5) is used to obtain the equation in terms of the unknowns and corresponding boundary conditions.

$$b\int_0^L \int_{-h/2}^{+h/2} (\sigma_x \delta\varepsilon_x + \sigma_z \delta\varepsilon_z + \tau_{xz}\delta\gamma_{xz})dzdx - \int_0^L q\delta w dx + \rho\int_{-h/2}^{+h/2}\int_0^L \left(\frac{\partial^2 u}{\partial t^2}\delta u + \frac{\partial^2 w}{\partial t^2}\delta w\right)dxdz = 0 \tag{5}$$

where ρ is the mass density. Substituting the strains obtained from Eq. (2) in Eq. (5) and integrating it by parts following governing equations are obtained.

$$\delta u_0 : A_{11}\frac{\partial^2 u_0}{\partial x^2} - B_{11}\frac{\partial^3 w_0}{\partial x^3} + C_{11}\frac{\partial^2 \phi_x}{\partial x^2} + D_{11}\frac{\partial^2 \psi_x}{\partial x^2} + I_{13}\frac{\partial \phi_z}{\partial x} + J_{13}\frac{\partial \psi_z}{\partial x} + A_I\frac{\partial^2 u_0}{\partial t^2}$$
$$- B_I\frac{\partial^3 w_0}{\partial x \partial t^2} + C_I\frac{\partial^2 \phi_x}{\partial t^2} + D_I\frac{\partial^2 \psi_x}{\partial t^2} = 0 \quad (6)$$

$$\delta w_0 : B_{11}\frac{\partial^3 u_0}{\partial x^3} - A_{s11}\frac{\partial^4 w_0}{\partial x^4} + C_{s11}\frac{\partial^3 \phi_x}{\partial x^3} + D_{s11}\frac{\partial^3 \psi_x}{\partial x^3} + I_{s13}\frac{\partial^2 \phi_z}{\partial x^2} + J_{s13}\frac{\partial^2 \psi_x}{\partial x^2} + B_I\frac{\partial^3 u_0}{\partial x \partial t^2}$$
$$- A_{SI}\frac{\partial^4 w_0}{\partial x^2 \partial t^2} + C_{SI}\frac{\partial^3 \phi_x}{\partial x^2 \partial t} + D_{SI}\frac{\partial^3 \psi_x}{\partial x \partial t^2} + A_I\frac{\partial^2 w_0}{\partial t^2} + C_{S1I}\frac{\partial^2 \phi_z}{\partial t^2} + D_{S1I}\frac{\partial^2 \psi_z}{\partial t^2} = 0 \quad (7)$$

$$\delta \phi_x : C_{11}\frac{\partial^2 u_0}{\partial x^2} - C_{s11}\frac{\partial^3 w_0}{\partial x^3} + C_{ss111}\frac{\partial^2 \phi_x}{\partial x^2} + C_{ss211}\frac{\partial^2 \psi_x}{\partial x^2} + I_{ss113}\frac{\partial \phi_z}{\partial x} + J_{ss113}\frac{\partial \psi_z}{\partial x}$$
$$- C_{sss155}\phi_x - C_{sss155}\frac{\partial \phi_z}{\partial x} - C_{sss255}\psi_x - C_{sss255}\frac{\partial \psi_z}{\partial x} + C_I\frac{\partial^2 u_0}{\partial t^2} - C_{SI}\frac{\partial^3 w_0}{\partial x \partial t^2} + C_{SS1I}\frac{\partial^2 \phi_x}{\partial t^2} \quad (8)$$
$$+ C_{SS2I}\frac{\partial^2 \psi_x}{\partial t^2} = 0$$

$$\delta \psi_x : D_{11}\frac{\partial^2 u_0}{\partial x^2} - D_{s11}\frac{\partial^3 w_0}{\partial x^3} + C_{ss211}\frac{\partial^2 \phi_x}{\partial x^2} + D_{ss111}\frac{\partial^2 \psi_x}{\partial x^2} + I_{ss213}\frac{\partial \phi_z}{\partial x} + J_{ss213}\frac{d\psi_z}{dx} - C_{sss255}\phi_x$$
$$- C_{sss255}\frac{\partial \phi_z}{\partial x} - D_{sss155}\psi_x - D_{sss155}\frac{\partial \psi_z}{\partial x} + D_I\frac{\partial^2 u_0}{\partial t^2} - D_{SI}\frac{\partial^3 w_0}{\partial x \partial t^2} + C_{SS2I}\frac{\partial^2 \phi_x}{\partial t^2} \quad (9)$$
$$+ D_{SS1I}\frac{\partial^2 \psi_x}{\partial t^2} = 0$$

$$\delta \phi_z : C_{sss155}\frac{\partial \phi_x}{\partial x} + C_{sss155}\frac{\partial^2 \phi_z}{\partial x^2} + C_{sss255}\frac{\partial \psi_x}{\partial x} + C_{sss255}\frac{\partial^2 \psi_z}{\partial x^2} - I_{13}\frac{\partial u_0}{\partial x} + I_{s13}\frac{\partial^2 w_0}{\partial x^2} - I_{ss113}\frac{\partial \phi_x}{\partial x} \quad (10)$$
$$- I_{ss213}\frac{\partial \psi_x}{\partial x} - I_{sss133}\phi_z - I_{sss233}\psi_z + C_{S1I}\frac{\partial^2 w_0}{\partial t^2} + C_{SS11I}\frac{\partial^2 \phi_z}{\partial t^2} + D_{SS12I}\frac{\partial^2 \psi_z}{\partial t^2} = 0$$

$$\delta \psi_z : C_{sss255}\frac{\partial \phi_x}{\partial x} + C_{sss255}\frac{\partial^2 \phi_z}{\partial x^2} + D_{sss155}\frac{\partial \psi_x}{\partial x} + D_{sss155}\frac{\partial^2 \psi_z}{\partial x^2} - J_{13}\frac{\partial u_0}{\partial x} + J_{s13}\frac{\partial^2 w_0}{\partial x^2}$$
$$- J_{ss113}\frac{\partial \phi_x}{\partial x} - J_{ss213}\frac{\partial \psi_x}{\partial x} - I_{sss233}\phi_z - J_{sss133}\psi_z + D_{S1I}\frac{\partial^2 w_0}{\partial t^2} + D_{SS12I}\frac{\partial^2 \phi_z}{\partial t^2} \quad (11)$$
$$+ D_{SS11I}\frac{\partial^2 \psi_z}{\partial t^2} = 0$$

where the stiffness coefficients used in the equations (6)-(11) can be obtained as

$$\left(A_{ij}, B_{ij}, A_{sij}\right) = Q_{ij}\int_{-h/2}^{+h/2}\left(1, z, z^2\right)dz,$$
$$\left(C_{ij}, C_{sij}, C_{ss1ij}, C_{ss2ij}, I_{ss1ij}, J_{ss1ij}\right) = Q_{ij}\int_{-h/2}^{+h/2} g_1(z)\left[1, z, g_1(z), g_2(z), g_1''(z), g_2''(z)\right]dz,$$
$$\left(D_{ij}, D_{sij}, D_{ss1ij}, I_{ss2ij}, J_{ss2ij}\right) = Q_{ij}\int_{-h/2}^{+h/2} g_2(z)\left[1, z, g_2(z), g_1''(z), g_2''(z)\right]dz,$$
$$\left(C_{sss1ij}, C_{sss2ij}\right) = Q_{ij}\int_{-h/2}^{+h/2} g_1'(z)\left[g_1'(z), g_2'(z)\right]dz, \left(D_{sss1ij}\right) = Q_{ij}\int_{-h/2}^{+h/2} g_2'(z)g_2'(z)dz,$$

$$\begin{aligned}
&\left(I_{ij}, I_{sij}, I_{sss1ij}, I_{sss2ij}\right) = Q_{ij} \int_{-h/2}^{+h/2} g_1''(z)\left[1, z, g_1''(z), g_2''(z)\right] dz,\\
&\left(A_I, B_I, A_{SI}\right) = \rho \int_{-h/2}^{+h/2} \left(1, z, z^2\right) dz,\\
&\left(C_I, C_{SI}, C_{SS1I}, C_{SS2I}\right) = \rho \int_{-h/2}^{+h/2} g_1(z)\left[1, z, g_1(z), g_2(z)\right] dz,\\
&\left(D_I, D_{SI}, D_{SS1I}\right) = \rho \int_{-h/2}^{+h/2} g_2(z)\left[1, z, g_2(z)\right] dz,\\
&\left(C_{CS1I}, C_{CSS11I}, D_{SS21I}\right) = \rho \int_{-h/2}^{+h/2} g_1'(z)\left[1, g_1'(z), g_2'(z)\right] dz,\\
&\left(D_{S1I}, D_{SS11I}\right) = \rho \int_{-h/2}^{+h/2} g_2'(z)\left[1, g_2'(z)\right] dz,\\
&\text{where } g_1(z) = \left(z - \frac{4z^3}{3h^2}\right), g_2(z) = \left(z - \frac{16z^5}{5h^4}\right)
\end{aligned} \tag{12}$$

g_1', g_1'' etc are the derivatives of the functions wrt z

The boundary conditions are (along x=0 to x=L)

$$\begin{aligned}
&N_x = 0 \text{ or } u_0 = 0; \quad M_x^b = 0 \text{ or } \partial w_0 / \partial x = 0; \quad \partial M_x^b / \partial x = 0 \text{ or } w_0 = 0;\\
&M_x^{s_1} = 0 \text{ or } \phi_x = 0 M_x^{s_2} = 0 \text{ or } \psi_x = 0; \quad Q_{xz}^1 = 0 \text{ or } \phi_z = 0; \quad Q_{xz}^2 = 0 \text{ or } \psi_z = 0
\end{aligned} \tag{13}$$

3. Navier's Solution Scheme

To find the unknown displacements, Navier's solution is used. To fulfil the above mentioned conditions at the boundary, the displacements and slopes are represented in the form of single trigonometric series as mentioned in Eq. (14) below.

$$\begin{aligned}
&\left(u_0, \phi_x, \psi_x\right) = \sum_{m=1}^{\infty} \left(u_m, \phi_{xm}, \psi_{xm}\right) \cos \alpha x \sin \omega t\\
&\left(w_0, \phi_z, \psi_z\right) = \sum_{m=1}^{\infty} \left(w_m, \phi_{zm}, \psi_{zm}\right) \sin \alpha x \sin \omega t\\
&\text{where} \left(\alpha = \frac{m\pi}{a}\right)
\end{aligned} \tag{14}$$

$u_m, w_m, \phi_{xm}, \psi_{xm}, \phi_{zm}$ and ψ_{zm} are the functions which are unknown. For vibration problem transverse and other mechanical loads are considered as zero. Substituting Eq. (14) in the governing equations (6)-(11) gives the following system of equations.

$$\left([K] - \omega^2 [M]\right)\{\Delta\} = \{0\} \tag{15}$$

$$[K] = \begin{bmatrix} \bar{K}_{11} & \bar{K}_{12} & \bar{K}_{13} & \bar{K}_{14} & \bar{K}_{15} & \bar{K}_{16} \\ \bar{K}_{12} & \bar{K}_{22} & \bar{K}_{23} & \bar{K}_{24} & \bar{K}_{25} & \bar{K}_{26} \\ \bar{K}_{13} & \bar{K}_{23} & \bar{K}_{33} & \bar{K}_{34} & \bar{K}_{35} & \bar{K}_{36} \\ \bar{K}_{14} & \bar{K}_{24} & \bar{K}_{34} & \bar{K}_{44} & \bar{K}_{45} & \bar{K}_{46} \\ \bar{K}_{15} & \bar{K}_{25} & \bar{K}_{35} & \bar{K}_{45} & \bar{K}_{55} & \bar{K}_{56} \\ \bar{K}_{16} & \bar{K}_{26} & \bar{K}_{36} & \bar{K}_{46} & \bar{K}_{56} & \bar{K}_{66} \end{bmatrix}, \{\Delta\} = \begin{Bmatrix} u_0 \\ w_0 \\ \phi_x \\ \psi_x \\ \phi_z \\ \psi_z \end{Bmatrix}, [M] = \begin{bmatrix} m_{11} & m_{12} & m_{13} & m_{14} & m_{15} & m_{16} \\ m_{12} & m_{22} & m_{23} & m_{24} & m_{25} & m_{26} \\ m_{13} & m_{23} & m_{33} & m_{34} & m_{35} & m_{36} \\ m_{14} & m_{24} & m_{34} & m_{44} & m_{45} & m_{46} \\ m_{15} & m_{25} & m_{35} & m_{45} & m_{55} & m_{56} \\ m_{16} & m_{26} & m_{36} & m_{46} & m_{56} & m_{66} \end{bmatrix} \tag{16}$$

where the element of stiffness and mass matrices are

$$\begin{aligned}
&\bar{K}_{11} = -A_{11}\alpha^2,\ \bar{K}_{12} = B_{11}\alpha^3,\ \bar{K}_{13} = -C_{11}\alpha^2,\ \bar{K}_{14} = -D_{11}\alpha^2;\ \bar{K}_{15} = I_{13}\alpha,\\
&\bar{K}_{16} = J_{13}\alpha, \bar{K}_{22} = -A_{s11}\alpha^4, \bar{K}_{23} = C_{s11}\alpha^3, \bar{K}_{24} = D_{s11}\alpha^3, \bar{K}_{25} = -I_{s13}\alpha^2,\\
&\bar{K}_{26} = -J_{s13}\alpha^2,\ \bar{K}_{33} = -C_{ss111}\alpha^2 - C_{sss155},\ \bar{K}_{34} = -\left[C_{ss211}\alpha^2 + C_{sss255}\right],\\
&\bar{K}_{35} = I_{ss113}\alpha - C_{sss155}\ \alpha, \bar{K}_{36} = J_{ss113}\alpha - C_{sss255}\alpha, \bar{K}_{45} = I_{ss213}\alpha - C_{sss255}\alpha,\\
&\bar{K}_{46} = J_{ss213}\alpha - D_{sss155}\alpha, \bar{K}_{56} = -C_{sss255}\alpha^2 - I_{sss233}, \bar{K}_{66} = -D_{sss155}\alpha^2 - J_{sss133}\\
&m_{11} = -A_I\omega^2,\ m_{12} = B_{11}\alpha\omega^2,\ m_{13} = -C_I\omega^2,\ m_{14} = -D_I\omega^2;\ m_{15} = 0,\\
&m_{16} = 0, m_{22} = -A_{SI}\alpha^2 - A_I\omega^2, m_{23} = C_{SI}\alpha\omega^2, m_{24} = D_{SI}\alpha\omega^2,\\
&m_{25} = -C_{S1I}\omega^2, m_{26} = -D_{S1I}\omega^2,\ m_{33} = -C_{SS1I}\omega^2,\ m_{34} = -C_{SS2I}\omega^2,\\
&m_{35} = 0, m_{36} = 0-, m_{45} = 0, m_{46} = 0, m_{56} = -D_{SS1I}\omega^2, m_{66} = -D_{SS1I}\omega^2
\end{aligned} \tag{17}$$

4. Results and Discussion

In this paper, the free vibration analysis of $0^0/90^0$ and $0^0/90^0/0^0$ laminated beams is performed. The beams are made of material having properties mentioned in Eq. (17) and the dimensionless form used for the determination of vibration frequencies are sated in Eq. (18).

$$\begin{aligned}
&E_1 / E_2 = \text{ Open}, E_3 = E_2, G_{12} = G_{13} = 0.6E_2, G_{23} = 0.5E_2,\\
&\mu_{12} = \mu_{13} = \mu_{23} = 0.25,\ \rho = \text{ constant}
\end{aligned} \tag{18}$$

$$\bar{\omega} = \frac{wL}{h}\sqrt{\left(\frac{\rho}{E_3}\right)} \tag{19}$$

Table 1 shows the results obtained using present FOSNDT and it's comparison with the results documented by Sayyad and Ghugal [6].

Table 1. Comparison of dimensionless vibration frequencies for different lamination schemes

Lamination Pattern	E_1/E_2	Source	*a/h* 5	10	20	50	100
$0^0/90^0$	40	Present	6.089	6.9704	7.267	7.358	7.372
		Sayyad and Ghugal [6]	6.185	6.999	7.276	7.377	7.413
		Khdeir and Reddy [13]	6.128	6.945	----	----	----
		Aydogdu [14]	6.144	----	7.218	----	----
		Vo and Thai [15]	6.058	6.909	7.204	7.296	----
		Vo et al. [16]	6.090	6.918	7.207	----	----
		Matsunaga [17]	6.144	6.991	----	----	----
		Timoshenko [3]	5.953	6.882	----	----	----
		Bernoulli–Euler [1,2]	7.124	7.269	----	----	----
$0^0/90^0/0^0$	40	Present	9.099	13.601	16.339	17.461	17.641
		Sayyad and Ghugal [6]	9.245	13.675	16.450	17.638	17.647
		Aydogdu [18]	9.168	13.552	16.308	17.455	17.641
		Reddy [5]	9.252	13.624	16.339	17.463	17.643
		Karama et al. [19]	9.208	13.614	16.337	17.462	17.643
		Khdeir and Reddy [13]	9.208	13.614	----	----	----
		Vo and Thai [15]	9.206	13.607	16.327	17.449	----
		Chalak et al. [20]	9.208	13.611	16.336	----	----
		Aydogdu [14]	9.207	----	16.337	----	----
		Vo et al. [16]	9.294	13.616	16.326	----	----
		Matsunaga [17]	9.817	14.149	----	----	----
		Timoshenko [3]	9.205	13.670	----	----	----
		Bernoulli–Euler [1,2]	17.42	17.632	----	----	----

The comparison of results shows that, the present FOSNDT gives results very close to all the theories documented in the literature. Without using the exact solution, the effect of transverse normal deformation cannot be shown as the fact that the vibration frequency is a quantity with respect to the neutral axis.

5. Conflict of Interest

The authors state that the there is no conflict of interest in this paper.

6. References

7. Bernoulli, J: Curvatura laminae elasticae. Acta Eruditorum Lipsiae, 262-276 (1694).

8. Euler, L: Methodus inveniendi lineas curvas maximi minimive proprietate gaudentes. Apud Marcum-Michaelem Bousquet and Socia, Lausanne, Geneva, Switzerland, 1-322 (1744).

9. Timoshenko, S. P.: On the correction for shear of the differential equation for transverse vibration of prismatic bars. Philos Mag 46, 744-760 (1921)

10. Timoshenko, S. P.: On the transverse vibrations of bars of uniform cross-section. Philos Mag 43, 125-131 (1922)

11. Reddy, J. N.: A simple higher order theory for laminated composite plates. ASME Journal of Applied Mechanics 51, 745-752 (1984).

12. Sayyad, A. S., Ghugal, Y.M.: Effect of thickness stretching on the static deformations, natural frequencies and critical buckling loads of laminated composite and sandwich beams. Journal of Brazilian Society of Mechanical Sciences and Engineering 40, 296 (2018).

13. Garg, A., Chalak, H. D.: Novel higher order zigzag theory for analysis of laminated sandwich beams. Proceedings of the Institution of Mechanical Engineers, Part L: Journal of Materials: Design and Applications 235(1), 176-194 (2021).

14. Garg, A., Chalak, H. D., Zenkour, A. M., Belarbi, M.-O., Sahoo, R.: Bending and Free Vibration Analysis of Symmetric and Unsymmetric Functionally Graded CNT Reinforced Sandwich Beams Containing Soft Core. Thin Walled Structures 170, (2022).

15. Chalak, H. D., Zenkour, A. M., Garg, A.: Free Vibration and Modal Stress Analysis of FG CNTRC Beams Under Hygrothermal Conditions Using Zigzag Theory. Mechanics Based Design of Structures And Machines 49(8), 1-22 (2021).

16. Naik, N. S., Sayyad, A. S.: 2D analysis of laminated composite and sandwich plates using a new fifth-order plate theory, Latin American Journal of Solids and Structures 15(9), 1-27 (2018).

17. Naik, N. S., Sayyad, A. S.: An accurate computational model for thermal analysis of laminated composite and sandwich plates. Journal of Thermal Stresses 42(5), 559-579 (2019).

18. Naik, N. S., Sayyad, A. S.: Analysis of laminated plates subjected to mechanical and hygrothermal environmental loads using fifth-order shear and normal deformation theory. International Journal of Applied Mechanics 12(3), 2050028(51pages) (2020).

19. Khdeir, A. A., Reddy, J. N.: Free vibration of cross-ply laminated beams with arbitrary boundary conditions. International Journal of Engineering Science 32(12),1971-1980 (1994).

20. Aydogdu, M.: Vibration analysis of cross-ply laminated beams with general boundary conditions by Ritz method. Internal Journal of Mechanical Sciences 47, 1740-1755 (2005).

21. Vo, T. P., Thai, H. T.: Vibration and buckling of composite beams using refined shear deformation theory. International Journal of Mechanical Sciences 62, 67-76 (2012).

22. Vo, T. P., Thai, H. T., Inam, F.: Axial-flexural coupled vibration and buckling of composite beams using sinusoidal shear deformation theory. Archive of Applied Mechanics 83(4), 605-622 (2013).

23. Matsunaga, H.: Vibration and buckling of multilayered composite beams according to higher order deformation theories. Journal of Sound and Vibration 246(1), 47-62(2001).

24. Aydogdu, M.: A new shear deformation theory for laminated composite plates. Composite Structures 89, 94-101 (2009).

25. Karama, M., Afaq, K. S., Mistou, S.: Mechanical behaviour of laminated composite beam by new multi-layered laminated composite structures model with transverse shear stress continuity. International Journal of Solids and Structures 40, 1525-1546 (2003).

26. Chalak, H. D., Chakrabarti, A., Iqbal, M. A., Sheikh, A. H.: Vibration of laminated sandwich beams having soft core. Journal of Vibration and Control 18(10), 1422-1435(2011).

A four node quadrilateral shell element for free vibration response of functionally graded spherical shell panels

R B Dahele[1], S D Kulkarni[2], V A Dagade[1]

[1]Research Scholar, [2]Former Associate Professor
Department of Civil Engineering, College of Engineering Pune, Pune 411005, Maharashtra, India.
rajendra2299@gmail.com

Abstract. The present work involves the free vibration analysis of functionally graded spherical shells using a four-node quadrilateral shell element. This four-node, shell element has seven DOF (Degrees of Freedom) per node namely; three displacements, two rotations of mid-plane, and two transverse shear strain com-ponents. The element is developed by modifying the discrete Kirchhoff quadri-lateral shell element based on Reddy's third-order theory developed earlier by the authors for the cylindrical shell. By transforming all the degrees of freedom from local to global coordinates using a transformation matrix converts the plate bending element into a flat shell element, resulting in the present formulation being suitable to give results for thin shells as well as for thick shells. In the present work functionally graded spherical shell panels with simply supported as well as clamped boundary condition, with various radius to span ratios and with different volume fraction indices are analyzed for free vibration response. The power law property variation through the thickness is considered in this study. To assess the performance of the developed element, the results, of non-dimen-sionalised frequencies are compared with the results presented in the literature. Comparison of results shows that the developed element yields quite accurate results even with the coarser mesh, which indicates the computational efficiency. It is also seen that the percentage difference between the present results and the other results available in the literature is less than 2 in most of the cases.
Keywords- *Functionally Graded Material, Spherical Shell, Free Vibration, Finite Element, Third-order.*

1. Introduction

The Functionally graded materials are useful for structural elements that are subjected to high differential temperatures. Laminated composite can also be used for the above applications, but at the layer interface, there is discontinuity of mechanical properties and are susceptible to delamination. FG materials do not suffer from delamination as the material properties vary smoothly across the thickness and thus advantages over the laminated composite materials. FG material is microscopically heterogeneous and usually contains two material phases namely; Ceramic and metal. Ceramic acts as a thermal barrier and metal impart the required ductility. In the literature, various 2D and 3D solutions have been presented by various researchers. In the present study, a four-node quadrilateral shell element is developed based on Reddy's third-order theory which was used earlier by for free vibration response of functionally graded cylindrical shells, since the finite element technique is useful for the analysis of spherical shells with various displacements-based boundary conditions.

Earlier, Dahale et.al. [1] have developed a four node quadrilateral element based on Reddy's theory for the free vibration analysis of functionally graded cylindrical shell panels, where C^1 continuity problem posed by Reddy's third order theory in finite element formulation is successfully circumvented by using improved discrete Kirchhoff interpolation functions. The Present finite element results are compared with the finite element results based on Sander's theory and Donnell's theory of Pandey et.al. [3] and Liu et.al. [4] The present results are also compared with the finite element results based on higher order shear deformation theory with eight-node and nine degrees of freedom per node of Pradyumna et.al. [5] It is seen form the literature that no researcher has so far used the four node quadrilateral plate element using the novel concept of obtaining the transformation matrix, which is used in present formulation for the analysis of functionally graded spherical shell panels.

The present element is observed to yield quite accurate results. The use of two separate coordinate systems for translational and rotational degrees of freedom and the use of the plate theory is the novelty of the present approach.

1.1. Displacement Approximation for the Reddy's Third Field Order Theory

Consider a spherical shell panel as shown in 'Figure 1' with thickness 'h'. The shell's middle surface is known to be a reference plane where z is zero and the top surface is at z = 'h/2'and bottom surface is at z = '-h/2'.

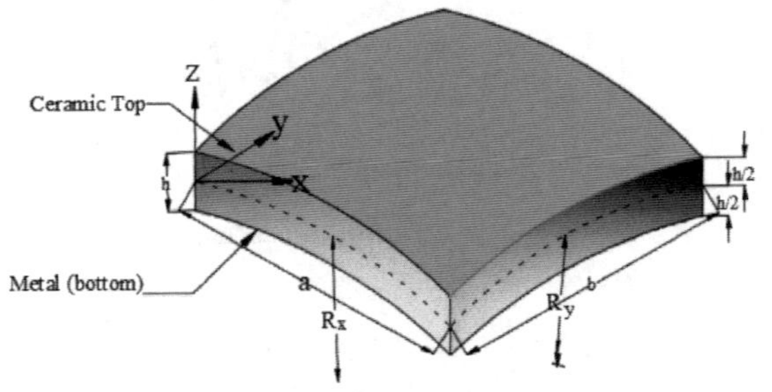

Figure 1. Geometry of FGM spherical shells.

The displacement field approximation of Reddy's third order theory assumes a third order variation for the in-plane displacement and a uniform transverse displacement across the thickness. The condition of zero shear stress at the top and bottom of the plate imposed and has same number of primary variables as the first order shear deformation theory. Reddy's third order theory does not require the shear correction factor unlike first order shear deformation theory. This theory can yield quite accurate results for thick as well as thin plates.

The displacement field approximation is as follows:

$$w(x,y,z,t) = w_0(x,y,t) \tag{1}$$

$$u(x,y,z,t) = u_0(x,y,t) - z\,w_{0d} + R(z)\varphi_0(x,y,t) \tag{2}$$

Where, $$\mathrm{u} = \begin{bmatrix} u_x \\ u_y \end{bmatrix}, \quad w_{0d} = \begin{bmatrix} w_{0,x} \\ w_{0,y} \end{bmatrix}, \quad u_0 = \begin{bmatrix} u_{0x} \\ u_{0y} \end{bmatrix}, \quad \varphi_0 = \begin{bmatrix} \varphi_{0x} \\ \varphi_{0y} \end{bmatrix} \tag{3}$$

and $$R(z) = \left[z - \frac{4z^2}{3h^2} \right]$$

1.2. Material Properties of FGM

Consider a spherical shell panel as shown in 'Figure 1'. The properties changes gradually from ceramic phase to the metal phase in functionally graded material. Voigt's rule of mixtures (ROM) is used for predicting the material property variation, through the thickness as follows.

$$P_z = P_b + \left(P_t - P_b\right)\left(0.5 \pm \frac{z}{h}\right)^n \tag{4}$$

In above equation n is the power law index which can vary from 0 to ∞. P_b represents the properties of the material at bottom and that of the top of the shell is P_t. P_z is the properties of the material at any z-coordinates. Poisson's ratio is assumed to be constant.

1.3. Finite Element Formulation

The spherical shell shown in 'Figure 1' is discretized using the quadrilateral elements developed by the same authors and presented in Dahale et.al. [1], with necessary modifications, as shown in 'Figure 2'. The actions and displacements are transformed from local to global coordinate system by using the concept of two co-ordinates systems as given by Clough et.al. [2].The orientations of local, global, and surface coordinate axes are shown in 'Figure 2'.

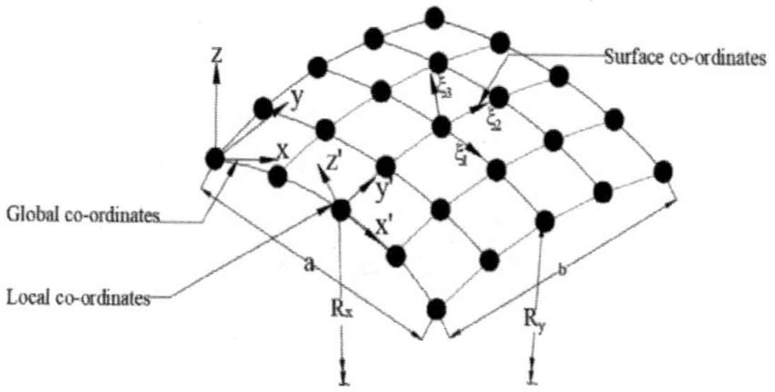

Figure 2. Discretized spherical shells with coordinate systems.

Cartesian coordinates (x, y, z) and surface coordinates (ξ_1, ξ_2, ξ_3) are shown in 'Figure 2'. In surface coordinate ξ3 is taken normal to the shell surface at every node of the element. It is assumed that rotation about the normal to the tangent plane would be negligible in the actual shell. Accordingly, rotational degrees of freedom referred to the surface coordinates (ξ_1, ξ_2) per nodal point are only included in the base system describing the assembled structure by Clough et.al. [2]". Thus in this approach, the local translational degrees $u_{0x}{}'', u_{0y}{}'', w_0{}''$ are transformed to base coordinate system with reference to global Cartesian coordinates (x, y, z) and the rotational degrees $\theta_{0x}{}'', \theta_{0y}{}'', \psi_{0x}{}'', \psi_{0y}{}''$ are transformed to surface coordinate system (ξ_1, ξ_2, ξ_3), neglecting the contribution to (ξ_3).
$U_i^{e'}$ is local and U_i^e is the global displacement vector for ith node shown in the following expressions.

$$U^{e'} = \begin{bmatrix} u_{0x}{}''^{i} & u_{0y}{}''^{i} & w_0{}''^{i} & \theta_{0x}{}''^{i} & \theta_{0y}{}''^{i} & \psi_{0x}{}''^{i} & \psi_{0y}{}''^{i} \end{bmatrix}^T \tag{5}$$

$$U^{e} = \begin{bmatrix} u_{0x}{}'^{i} & u_{0y}{}'^{i} & w_0{}'^{i} & \theta_{0\xi1}{}'^{i} & \theta_{0\xi2}{}'^{i} & \psi_{0\xi1}{}'^{i} & \psi_{0\xi2}{}'^{i} \end{bmatrix}^T \tag{6}$$

Transformation matrix T^i for transforming the local translational DOF to global translational DOF for an ith node is given by the following expression.

$$\begin{bmatrix} u_{0x}'^{i} \\ u_{0y}'^{i} \\ w_0'^{i} \end{bmatrix} = \begin{bmatrix} \lambda_{11} & \lambda_{12} & \lambda_{13} \\ \lambda_{21} & \lambda_{22} & \lambda_{23} \\ \lambda_{31} & \lambda_{32} & \lambda_{33} \end{bmatrix} \begin{bmatrix} u_{0x}^{i} \\ u_{0y}^{i} \\ w_0^{i} \end{bmatrix}$$

$$i.e. \quad u_0'^{i} = T_i u_0^{i} \tag{7}$$

where λ_{11}, λ_{12} etc. are the direction cosines of the local (primed) axes with respect to global (unprimed) axes.

The relationship between the local rotations and global rotations for an ith node is given by the following expression.

$$\begin{bmatrix} \theta_{0x}'^{i} \\ \theta_{0y}'^{i} \\ \theta_{0z}'^{i} \end{bmatrix} = \begin{bmatrix} \lambda_{22} & \lambda_{21} & \lambda_{23} \\ \lambda_{12} & \lambda_{11} & \lambda_{13} \\ \lambda_{32} & \lambda_{31} & \lambda_{33} \end{bmatrix} \begin{bmatrix} \theta_{0x}^{i} \\ \theta_{0y}^{i} \\ \theta_{0z}^{i} \end{bmatrix}$$

$$i.e. \quad \theta_0'^{i} = \overline{T}_i \theta_0^{i} \tag{8}$$

The relationship between the surface rotations $\theta_{0\xi}^{i}$ and θ_0^{i}, for the ith node can be obtained as follows, For a spherical shell, the angle between the global x-axis and tangent ξ_1 can be obtained easily as this tangent is actually tangent to the circle $x^2 + z^2 = R^2$ in the x-z plane. Differentiating this expression for the circle yields, $\tan\alpha = -x/z$. α is the angle between the axes ξ_1 and x. In addition, as the axis ξ_2 is along the global y-axis, is the tangent to the circle $y^2 + z^2 = R^2$ in y-z plane, considering the angle between the global y axis and ξ_2 is β i.e. $\tan\beta = -y/z$. Using this, the relationship between $\theta_{0\xi}^{\ i}$ and $\theta_0^{\ i}$ for the ith node can be obtained. Transformation matrix $\overline{T_\xi^i}$ for transforming the surface rotational DOF to global rotational DOF for i^{th}node is given by the following expression.

$$\begin{bmatrix} \theta_{0\xi_1}^{\ i} \\ \theta_{0\xi_2}^{\ i} \\ \theta_{0\xi_3}^{\ i} \end{bmatrix} = \begin{bmatrix} \lambda_{11} & \lambda_{12} & \lambda_{13} \\ \lambda_{21} & \lambda_{22} & \lambda_{23} \\ \lambda_{31} & \lambda_{32} & \lambda_{33} \end{bmatrix} \begin{bmatrix} \theta_{0x}^{\ i} \\ \theta_{0y}^{\ i} \\ \theta_{0z}^{\ i} \end{bmatrix} \tag{9}$$

$$i.e. \quad \theta_{0\xi}^{\ i} = \overline{T_\xi^i} \theta_0^{\ i}$$

Using eq. (8) and eq. (9), relationship between the $\theta_0'^{i}$ and $\theta_{0\xi}'^{i}$

$$\begin{aligned} \theta_{0\xi}^{\ i} &= T_\xi^{\ i} \overline{T}^{T^i} \theta_0'^{i} \\ &= \overline{T}_i^{\ \xi} \theta_0'^{i} \end{aligned} \tag{10}$$

The transformation matrix T_i^ξ of size 2×2, which relates the local rotations $\theta_{0x}'^{i}$, $\theta_{0y}'^{i}$ to tangential rotations $\theta_{0\xi_1}^{\ i}, \theta_{0\xi_2}^{\ i}$ are obtained by removing the third row and third column of transformation matrix $\overline{T_\xi}^{Ti}$. T_i^ξ is also used for transforming $\psi_{0x}'^{i}$, $\psi_{0y}'^{i}$, to tangential shell directions. Finally, the element transformation matrix of size 28×28 T^e is given as,

$$T^e = \begin{bmatrix} T_1 & 0 & 0 & 0 & 0 & 0 & 0 & 0 & 0 & 0 & 0 & 0 \\ 0 & T_1^\xi & 0 & 0 & 0 & 0 & 0 & 0 & 0 & 0 & 0 & 0 \\ 0 & 0 & T_1^\xi & 0 & 0 & 0 & 0 & 0 & 0 & 0 & 0 & 0 \\ 0 & 0 & 0 & T_2 & 0 & 0 & 0 & 0 & 0 & 0 & 0 & 0 \\ 0 & 0 & 0 & 0 & T_2^\xi & 0 & 0 & 0 & 0 & 0 & 0 & 0 \\ 0 & 0 & 0 & 0 & 0 & T_2^\xi & 0 & 0 & 0 & 0 & 0 & 0 \\ 0 & 0 & 0 & 0 & 0 & 0 & T_3 & 0 & 0 & 0 & 0 & 0 \\ 0 & 0 & 0 & 0 & 0 & 0 & 0 & T_3^\xi & 0 & 0 & 0 & 0 \\ 0 & 0 & 0 & 0 & 0 & 0 & 0 & 0 & T_3^\xi & 0 & 0 & 0 \\ 0 & 0 & 0 & 0 & 0 & 0 & 0 & 0 & 0 & T_4 & 0 & 0 \\ 0 & 0 & 0 & 0 & 0 & 0 & 0 & 0 & 0 & 0 & T_4^\xi & 0 \\ 0 & 0 & 0 & 0 & 0 & 0 & 0 & 0 & 0 & 0 & 0 & T_4^\xi \end{bmatrix} \quad (11)$$

After deriving the expressions for transformation matrix T^e, local stiffness matrix $K^{e'}$, local mass matrix $M^{e'}$and local load vector $P^{e'}$, finally the element stiffness matrix K^e of size 28×28, element mass matrix M^e of size 28 × 28 and element load vector P^e of size 28×1 with reference to global axes are given as

$$\begin{aligned} M^e &= T^{e^T} M^{e'} T^e, \\ K^e &= T^{e^T} K^{e'} T^e, \\ P^e &= T^{e^T} P^{e'} \end{aligned} \quad (12)$$

1.4. Assembly of element matrices

Summing up the contributions for all elements Hamilton's principle yields

$$\bar{M}\ddot{U} + \bar{K}U = \bar{P} \quad (13)$$

where $\bar{K}, \bar{M}$, *and* $\bar{P}$ are the assembled counterparts of matrices K^e , M^e, and P^e. For synchronous vibration, the $\bar{P}$ (load vector) is set to zero. The undamped natural frequencies ω_n are obtained by solving eigenvalue problem.

2. Numerical Results

A computer program is developed in FORTRAN for the free vibration analysis based on the formulation in section 2. A computer program and finite element formulation are assessed by comparing the result of non-dimensionalised natural frequencies obtained using developed element with the results in the literature. In all the problems considered in the study, the full shell panel is discretized with N × N mesh of equal size elements. The shell type and material properties considered are specified in 'Table 1'.

Table 1. Material Properties.

Shell type	Material	Modulus of Elasticity GPa	Density kg/m^3	Poisson's ratio
FGM1 (top ceramic)	Ceramic - Silicon carbide	$E_c = 427$	$\rho_c = 3100$	$\upsilon_c = 0.3$
	Metal - Aluminum	$E_m = 70$	$\rho_m = 2707$	$\upsilon_m = 0.3$
FGM2 (top ceramic)	Ceramic - Alumina	$E_c = 380$	$\rho_c = 3000$	$\upsilon_c = 0.3$
	Metal - Aluminum	$E_m = 70$	$\rho_m = 2707$	$\upsilon_m = 0.3$

The obtained natural frequencies are non-dimensionalised as-

- FGM1 $$\bar{\omega} = \omega \frac{a^2}{h}\sqrt{\frac{\rho_m}{E_m}},$$ (14)

- FGM2 $$\bar{\omega} = \omega a^2 \sqrt{\frac{\rho_m h}{D_m}}, \quad where\ D_m = \frac{E_m h^3}{(12(1-\upsilon_m^2))}$$ (15)

2.1. All-round simply supported spherical shell panel with square plan-form

The results of the natural frequencies of fundamental mode are presented for a spherical shell panel for 'a/h' = 5, 10, and 20 for two different 'R/a' ratios. The present results are compared with the results of Pandey et.al. [3] obtained using eight-node element with nine degrees of freedom per node shell element based on Sander's theory and Donnell's shell theories. In the analysis of the shell with mesh size 8 × 8 gives total 225 nodes and 2025 DOF. The results of Liu et.al. [4] obtained using four-node with nine degrees of freedom per node shell element based on the same theories are also given in the table for comparison. In the analysis of the shell with mesh size 8 × 8 gives total 81 nodes and 729 DOF. It is observed that for thick to thin shell panels the present element results are quite comparable with those of Pandey et.al. [3] and Liu et.al. [4]. It is noted that the present element is not based on shell theory. It is also observed that even though the element is based on plate theory it gives sufficiently accurate results not only for the shallow shell but for deep shells too. The 24 ×24 mesh size shell having 625 total nodes and 4375 total degrees of freedom. It is also to be noted that as the volume fraction increases, the values of the non-dimensionalised frequencies decrease because of the increase in metal part and thereby reducing the stiffness.

Table 2. Non-dimensionalised frequencies of all-round simply supported FGM1 spherical shell.

R/a	a/h	*n*	Pandey et. al. [3]		Liu et. al. [4]		DKTOTSS-FG (Present)		
			Sander's [3] 8 × 8	Donnell's [3] 8 × 8	Sander's [4] 8 × 8	Donnell's [4] 8 × 8	12 ×12	16 ×16	24 ×24
2	5	0.5	10.8079	10.3851	10.5452	10.7312	11.0092	11.0076	10.9861
		1	9.5173	9.2030	9.2754	9.4626	9.7803	9.7794	9.7584
		2	8.3062	8.0760	8.1172	8.3004	8.6135	8.6083	8.5911
		5	7.3351	7.1920	7.2739	7.4478	7.6106	7.6022	7.5903
	10	1	12.6834	12.5339	12.4847	12.6192	13.2238	13.1906	13.1887
		5	9.7248	9.6171	9.6718	9.7993	10.2062	10.1930	10.1801
	20	1	20.0728	20.0347	19.9198	19.9947	20.8099	20.7455	20.6781
		5	15.0152	14.9544	15.0160	15.0879	15.6747	15.6249	15.6420
5	5	0.5	10.0214	9.6165	9.9333	9.9814	10.1076	10.1063	10.1061
		1	8.8273	8.5231	8.7506	8.8049	8.8925	8.8848	8.8804
		2	7.7408	7.5318	7.6966	7.7533	7.7825	7.7712	7.7643
		5	6.9220	6.8212	6.9912	7.0442	6.9508	6.9424	6.9374
	10	1	9.9609	9.7961	9.8998	9.9405	10.1568	10.1424	10.1334
		5	7.9364	7.8511	7.9657	8.0054	8.0840	8.0690	8.0595
	20	1	11.8704	11.8198	11.8033	11.8306	12.1710	12.1461	12.1297
		5	9.2866	9.2509	9.3239	9.3503	9.5714	9.5470	9.5307

The non-dimensionalised fundamental frequencies of FGM2 spherical shells are analyzed in 'Table 3' for 'R/a' = 1, 5, and 10 with 'a/h' ratio is equal to 10. The shells of the power law indices '*n*' varying from 0, 0.2, 0.5 and 1 are presented in the table. The present results obtained using 12 × 12, 16 ×16, and 24 × 24 mesh are compared with those presented by Pradyumna et.al. [5] obtained using eight-node element having nine degrees of freedom per node. The formulation is based on the higher-order shear deformation theory. It is observed that for deep to shallow shell panels the present element results are quite comparable with those of Pradyumna et.al. [5].

Table 3. Non-dimensionalised frequencies of all-round simply supported FGM2 spherical shell.

R/a	*n* →	*Mesh size*	*0*	*0.2*	*0.5*	*1*
1	Pradyumna [5]	8×8	78.2306	72.6343	66.5025	59.8521
	DKTOTSS-FG	12×12	82.6259	76.7050	70.2166	63.2263
		16×16	81.8414	75.9609	69.5103	62.5522
		24×24	80.9123	75.1218	68.7532	61.8604
5	Pradyumna [5]	8×8	44.0073	41.7782	38.7731	34.6004
	DKTOTSS-FG	12×12	43.8301	42.0462	37.8716	33.6194
		16×16	43.8747	42.0790	37.8763	33.6161
		24×24	43.9458	42.1399	37.9145	33.5979
10	Pradyumna [5]	8×8	42.3579	40.2608	37.3785	33.3080
	DKTOTSS-FG	12×12	42.1685	40.4163	36.3157	32.1643
		16×16	42.1953	40.4358	36.3156	32.1430
		24×24	42.2171	40.4522	36.3176	32.1299

2.2. All-round clamped spherical shell panel with square plan-form

The fundamental non-dimensionalised frequencies of all-round clamped square planform FGM1 spherical shells for 'a/h' = 5, 10, and 20 for two different 'R/a' ratios are presented in 'Table 4'. The results are compared with Liu et.al. [4]. The percentage difference between Liu et.al. [4] and present element using 24 × 24 mesh size are given in the table. It is noted from 'Table 4', that the results for this case also are in good agreement with those given by Liu et.al. [4].

Table 4. Non-dimensionalised frequencies of all-round clamped FGM1 spherical shell.

R/a	a/h	n	Liu et. al. [4]	DKTOTSS-FG			%
			8 × 8	12 × 12	16 × 16	24 × 24	diff.
2	5	0.5	16.9465	17.0536	17.0003	16.9523	0.03
		1	15.0375	15.2261	15.1762	15.1322	0.63
		2	13.1326	13.3571	13.3080	13.2660	1.02
		5	11.3148	11.5103	11.4609	11.4191	0.92
	10	1	20.1030	20.5908	20.5843	20.5760	2.35
		5	15.6096	15.9336	15.9174	15.9019	1.87
	20	1	29.2643	30.3452	30.3793	30.4040	3.89
		5	22.5403	23.1489	23.1696	23.1844	2.86
5	5	0.5	15.9159	15.9972	15.9294	15.8687	0.30
		1	14.1023	14.1574	14.0937	14.0382	0.45
		2	12.3124	12.3446	12.2826	12.2301	0.67
		5	10.6242	10.6858	10.6247	10.3877	2.23
	10	1	16.9965	17.2719	17.2433	17.2161	1.29
		5	13.4081	13.5671	13.5305	13.2578	1.12
	20	1	19.6830	20.2246	20.2254	20.2244	2.75
		5	15.6730	15.9864	15.9772	15.6595	0.09

The results for the first six mode shapes for FGM1 shallow shell panel with 'R/a' = 5 and 'a/h' = 5 for volume fraction index = 1 are plotted in Fig 4 and Fig 5 for the all-round clamped and for one edge clamped and other three free (cantilever) boundary conditions respectively. The mesh size of 24 × 24 is used for obtaining mode shapes.These results will serves as a benchmark for the other researchers.

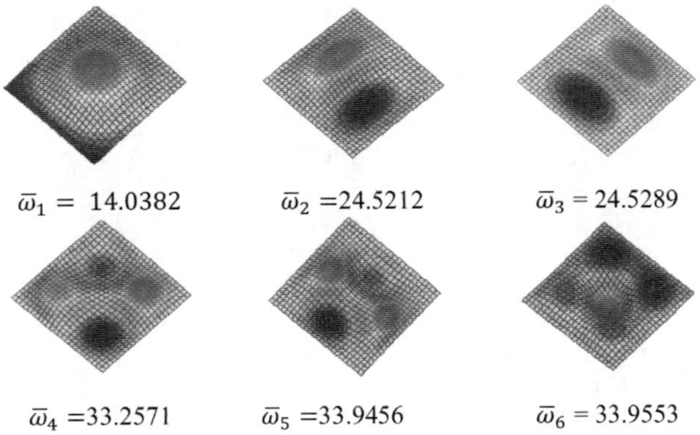

Fig 4. First six mode shapes of all-round clamped FGM1 shells for 'R/a'=5, 'a/h'=5 'n' = 1.

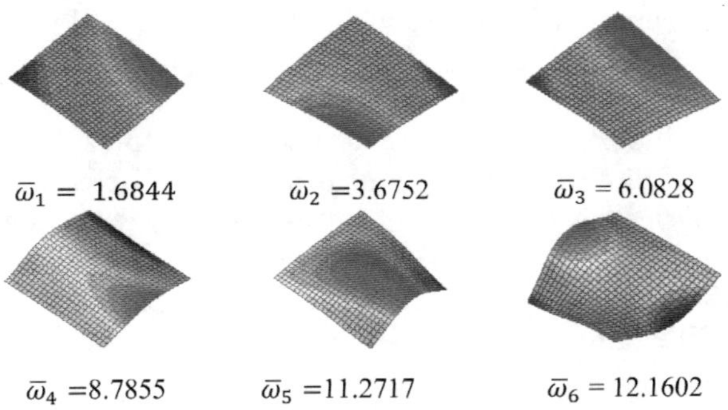

Fig 5. First six mode shapes of cantilever FGM1 spherical shell for 'R/a'=5, 'a/h'=5, 'n'= 1.

The non-dimensionalised frequencies of the all-round clamped FGM2 spherical shells are analyzed in 'Table 5' for 'R/a' = 1, 5, and 10 and 'a/h' ratio of 10. The present results obtained are compared with those presented by Pradyumna et.al. [5]. It is observed that for deep to shallow shell panels the present element results are quite comparable with those of Pradyumna et.al. [5].

Table 5. Non-dimensionalised frequencies of all-round clamped FGM2 spherical shell.

R/a	$n \rightarrow$	Mesh size	0	0.2	0.5	1
1	Pradyumna [5]	8 × 8	120.9210	112.2017	102.5983	92.2147
	DKTOTSS-FG	12 × 12	122.1768	113.9150	104.6969	94.6296
		16 × 16	122.2951	114.1103	104.8073	94.7293
		24 × 24	122.3743	114.1103	104.8850	94.7997
		% diff.	1.20	1.70	2.23	2.80
5	Pradyumna [5]	8 × 8	73.5550	69.6597	64.6114	57.8619
	DKTOTSS-FG	12 × 12	74.2272	71.2583	64.2484	57.0335
		16 × 16	74.1496	71.1737	64.1603	56.9321
		24 × 24	74.0600	71.0880	64.0772	56.8439
		% diff.	0.69	2.05	0.83	1.76
10	Pradyumna [5]	8 × 8	71.4659	67.7257	62.8299	56.2222
	DKTOTSS-FG	12 × 12	72.3846	69.4558	62.5596	55.4823
		16 × 16	72.2942	69.3684	62.4696	55.3791
		24 × 24	72.1926	69.2713	62.3759	55.2806
		% diff.	1.02	2.28	0.72	1.67

3. Conclusions

The present element gives sufficiently accurate results for natural frequencies for all cases of functionally graded shells including the shallow and deep shells considered in this study. The percentage difference between the present results and those reported in literature is even less than 0.5 in some cases and mostly below 2. It is also observed from the tables that the results obtained are converging as the mesh becomes fine. It is also observed that, the present formulation is suitable for capturing the modes, which the 2D analytical theory fails to capture. The element developed for free vibration response of functionally graded spherical shells are suitable for the general purpose finite element programming, and do not suffer from any locking problem, hence it is useful to thick as well as thin shell. This proves the suitability of the developed element for the functionally graded spherical shells. The use of the two coordinate system in the transformation of translational and rotational degrees of freedom from local to global system, is an important contribution of the present study.

References

1. Dahale R B, Kulkarni S D and Dagade V A 2021 Free Vibration Response of Functionally Graded Cylindrical Shells Using a Four Node Flat Shell Element. *Advances in construction technology and management-2021, International conference 2021*
2. Clough R W and Johnson CP 1968 A finite element approximation for the analysis of thin shells. *International Journal of Solids and Structures* 4 (1) 43-60
3. Pandey S and Pradyumna S 2015 A layerwise finite element formulation for free vibration analysis of functionally graded sandwich shells. *Composite Structures* vol 133, pp 438–450
4. Liu B, Ferreira A J M, Xing Y F and Neves A M A 2016 Analysis of functionally graded sandwich and laminated shells using a layerwise theory and a differential quadrature finite element method. *Composite Structures* vol 136, pp 546–553
5. Pradyumna S and Bandyopadhyay J N 2008 Free vibration analysis of functionally graded curved panels using a higher-order finite element formulation. *Journal of Sound and Vibration* vol 318, pp 176–192

Accident prediction modelling for expressways: a review

P Kumar[1], J K Jain[1] and G Singh[2]

[1]Civil Engineering Department, Malaviya National Institute of Technology, Jaipur, India
[2]Civil Engineering Department, Deenbandhu Chhotu Ram University of Science and Technology, Murthal, Sonepat, Haryana, India
{parveen.rh.kumar, jkjain05, singhgyan27}@gmail.com

Abstract. Expressways are the need for fast movement of goods and human being for long distances in current scenario for every country, but at the same time these facilities have the highest severity rate among all the road categories. Various researchers around the world have tried to find out the effect of traffic volume, road geometry and environmental factors on frequency of accident on expressway by using different accident prediction modeling methods during last three decades. The purpose of this review paper is to find out the appropriate modeling method which can be used to predict accident frequency on the expressways in the developing country like India where traffic conditions, vehicular characteristics, and the driver behaviour are very different from the developing countries and to find out the research areas where the findings are inconclusive. Literature review suggests that among various models used so far, correlated random parameter model found to be the most advanced model to simultaneously account for both the heterogeneous effects of explanatory factors across the road segments and the cross correlations among the random parameter estimates. Findings related to variables like percentage of heavy vehicles, vertical gradient and lane width are not conclusive. However, safety effect of variables like speed limit, roadway lighting, pavement type and fog are less studied on expressway. The effect of allowing two wheelers on the expressways safety has not been studied yet, and the accidents related to fatigue or drowsiness also need examination.
Keywords: Accident Frequency Models, Explanatory Variables, Review, Expressway.

1. Introduction

Road traffic accidents are the important cause of death and injury in all countries, causing more than 1.35 million deaths each year and about 50 million injuries globally. About ninety percent of these deaths and injuries are being happened in developing countries [1]. Among the developing countries, India have the highest numbers of traffic deaths in the world with 1,51,113 persons killed in 2019 [2]. Road traffic deaths and injuries become unnecessary burden on the health system and as well as onthe economy of the developing countries.

Road safety experts, researchers and various agencies related to road transportation all over the world are continuously trying to understand various factors affecting accident frequency and their severity outcomes so that proper countermeasures and policies can be adopted to reduce these unwanted incidences. In line with these efforts, accident prediction models (APMs) have been proved an important tool to develop statistical relationship or algorithm between accident frequency or severity and various independent variables like traffic volume, road geometry and road environ- mental factors etc. Literature suggests that selection of a particular type of model is based upon the requirement of the planners or engineers and also effected by various data and methodological issues like "over-dispersion, under-dispersion, time varying explanatory variables, temporal and spatial correlation, low sample mean and small sample size, injury-severity and accident-type correlation, under-reporting, omitted variables, endogenous variables, functional form and fixed parameters" [3] . Other than these, unobserved heterogeneity is also an important issue that affects the choice of APM [4]. These APMs can be divided broadly into parametric models and non- parametric models [5]. Parametric models have major advantage of quantifying the effects of different causal factors on road traffic accidents. However, at the same time these models require assumption of predefined relationship between accidents and their causal factors. On the other hand, new generation of non-parametric methods requires no such assumptions and their predictive power is much better than statistical models. A major limitation of non-parametric models is that these are not capable to quantifying the effects of contributory factors related to road traffic accidents [6].

Expressways are the controlled access highest class of roads being constructed for fast movement of people and goods from one place to another. About 1642 km length of expressway is in operation at present in India. However, Government of India planned to construct 23 more expressways of about 7800 km length in next five years. Expressways have highest accident severity rate due to their high speed limit and diversity in traffic composition as compared to other types of highway facility [7–12]. Similarly, in India also expressways have high accident severity rate as compared to national highways and state highways [8]. Some of the probable causes of this high accident severity rate on expressways in India have been reported as driver's fatigue and over-speeding [13], lack of foot over bridges for pedestrian crossing, non- availability of vehicle lay-bye and after crash care facility [14], practice of raised median instead of flushed or depressed median, non-standard guard rail installation, unsafe entry and exit ramps, non-standard crash cushions and chevron markings [15]. But none of the research tries to quantify effect of these factors on accidents.

Keeping the above in mind, this paper reviews more than 60 papers comprising different accident prediction modeling methods, and various explanatory variables used for the estimation of frequency of accidents on expressways all around the world to understand advantages and shortcomings of the adopted methods and to find out gaps in research work.

2. Literature Review

2.1. Parametric Accident Prediction Models

Accident is a discreet variable and a rare event. Hence, Poisson regression found as an appropriate method to model the accident frequency data on expressways as com- pared to linear models [16–20]. However, data related to accident frequency generally found to be over-dispersed, thus violating basic assumption of Poisson model (mean and variance are equal). So, Poisson model is not capable of handling over-dispersion and under-dispersion. Its results are seriously affected when size of the sample as well as the sample mean are small [3]. To overcome these limitations and enhance the predictive power of the Poisson model many variants of Poisson model are in use.

The zero-truncated Poisson (ZTP) model [21], also known as conditional Poisson distribution or the positive Poisson distribution, is used when due to limitations of observability and omission of certain accident data, the sample has to be truncated. Hosseinlou et al.[19] developed a zero-truncated Poisson (ZTP) model for the estimation of accidents and potential violations on Iran's expressway and compared the results with Poisson model. The results show that ZTP could fit data better than Poisson model for truncated sample.

The negative binomial/Poisson gamma (NB) model has been widely used during last three decades for expressways by various researchers. It is an extension of Poisson model [3], effectively dealing with over-dispersion issue of accident data. However, it has limitations in addressing the correlation in data due to temporal effects and location specific effects as it assumes accident frequency on a road segment as independent [7, 22–24].

Poisson lognormal model can account over-dispersion issue [25] and found more flexible than NB model [3]. However, it has limitation of complex model estimation. Caliendo et al.[17] developed negative multinomial model for expressway to address the problem of correlation among observations. Similar to Poisson and NB model, negative multinomial model also do not have the capacity of handling under- dispersion [3].

A univariate negative binomial (UVNB) model was developed by Li et al.[26] for the estimation of accident likelihood by collision type on the expressway in the state of Florida. Univariate or independent count NB models do not address correlation problem between different severity levels [26]. To address correlation problem, bivariate or multivariate approaches can be applied. However, these models are complex to estimate due to formulation of correlation matrix [3]. A bivariate negative binomial (BVNB) model was applied by Chen et al. [27] to analyze the effect of geometric elements on the safety of expressway. A multivariate negative binomial (MVNB) model was applied by Anastasopoulos [28] to estimate injury-severity frequency for expressway in Indiana.

A multivariate Poisson lognormal (MVPLN) model was estimated by Li et al. [26] to estimate the impact of independent variables simultaneously on various types of accidents, and the results showed that prediction accuracy of MVPLN model was found high as compared to the UVNB model. The MVPLN model can handle negative correlations, whereas MVNB model is not capable of handling this issue [29]. Wang et al. [30] was also recommended the use of the MVPLN in estimating weather related accidents by accident type and severity, and results showed more unbiased parameter estimates as compared to the UVPLN model.

Haq et al. [31] applied a zero inflated negative binomial (ZINB) model for the expressway in the state of Wyoming to develop APMs for truck accidents. The ZINB model is good for over-dispersed accident frequency data that have excess zeros and offers enhanced goodness-of-fit than the NB model [32]. This model assumes that there are two different state of accident generating process: an accident free state, and a normal accident prone state. However, the ZINB model does not correctly repro- duce the accident-data generating process [32]. Multivariate zero inflated negative binomial (MVZINB) model can handle correlation issue among different injury-severity levels and types of accident [28, 33].

The unobserved heterogeneity arising from temporal and spatial effects, results biases in parameter estimation in Poisson and NB model, as the accident distributions for similar road sections are considered the same in these models [34]. Many researchers [7, 35–38] have used random effect negative binomial (RENB) model to address the problem of unobserved heterogeneity while estimating accident frequency on expressways. The results showed that RENB model performed better as compared to NB model in handling unobserved heterogeneity issue. However, if the assumptions regarding variability of spatial and temporal effects are not correct, RENB model may be insufficient [37]. Some researchers [11, 18, 39] developed random effect model for expressways to address the unobserved variations at different sites. Zeng et al. [18] developed Random effect model with autoregression-1 (AR-1) correlation structure (REAR-1) to address chronological association among accidents occurred in the consecutive months and also unseen variations and temporal correlation present in the data.

Many researchers have used random parameter negative binomial (RPNB) model to address the issue of unobserved heterogeneity for the estimation of accidents on expressways [35, 37, 38, 40–43]. RPNB model is an extension of the RENB model [37], and have more flexibility in handling unobserved heterogeneity as compared to the later one. However, parameter estimation process for RPNB model is more complicated. The RPNB model cannot address the problem of probable correlation between the varied effects of explanatory factors [38]. Hou et al.[38] developed a correlated random parameter negative binomial (CRPNB) model for expressway tunnel safety in China and the results showed better goodness of fit as compared to RPNB model. CRPNB model estimates were found better accounting for cross correlations among the random-parameter estimates as well as heterogeneous effects of explanatory factors [43].

Due to the increasing availability of geo-referenced accident data and increasing computational power, Liu et al. [44] have estimated a geographically weighted negative binomial (GWNB) model to predict accident frequencies on expressway. GWNB model is capable of handling spatial heterogeneity and outperforms the traditional NB model. However, it does not address temporal heterogeneity. GWNB models are highly localized and they cannot be generalized to other locations [44].

Random parameters negative binomial-lindley (RPNB-L) model has been used by Rahman Shaon et al. [42] for handling expressway accident data with unobserved heterogeneity and high over-dispersion, and the results showed that this model outperformed RPNB model.

APMs for expressways have also been described in Highway Safety Manual (HSM) supplement [45]. These prediction models consist of a base safety performance function (SPF) with accident modification factors (AMFs) and a calibration factor to estimate the predicted average accident frequency. Using HSM approach La Torre et al. [9] analyzed single and multiple vehicle casualty and severe accidents for Italian expressways and provided a trustworthy tool for expressway agencies to deal with potential safety issues.

The generalized estimating equation (GEE) is an extension of generalized linear models to accommodate correlated dependent variable [46]. There are different approaches in this method to handle serial correlation including independent, dependent, autoregressive 1, exchangeable and unstructured correlation structure [3]. The GEE method was used by Mohammadi et al. [47] and Zheng et al. [46] to model the yearly accident counts on expressways to accommodate the temporal correlation problem.

To account for the effect of unobserved heterogeneity and spatio-temporal correlation in the data, Bayesian spatial and temporal approaches have also been proposed for the accident prediction [11].

Conditionally auto regressive (CAR) model is most common Bayesian spatial model, capable of handling correlation issue among the consecutive segment on the expressway [11]. Spatial Poisson-lognormal model is useful for addressing unobserved heterogeneity and spatial correlation among neighboring road segments of expressway [35].

2.2. *Non-Parametric Accident Prediction Models*

Non-parametric models do not require any prior assumption of the relationship between dependent variable and explanatory variables [20, 48]. Artificial Neural Network (ANN) model is one of the commonly used models suitable for nonlinear interactions among variables and it allows the inclusion of a large number of variables. However, ANN does not explain parameter estimates associated to the explanatory factors [20] and the results cannot be applied to the data sets with different specifications [49].

Classification and Regression Tree (CART) is a good tool for prediction and classification problems. This model can effectively handle collinearity and outliers problem which may be an important issue with regression models [50]. It was found a better substitute for NB regression models to analyze expressway accidents. However, CART model has its own limitations of not make use of continuous, ordinal variables, have problems in handling the interactions between independent variables and in doing elasticity analysis or sensitivity analysis [50].

2.3. *Significant Explanatory Variables*

The literature suggests that the choice of explanatory variables into the model has been based upon availability of data, significant variables established in earlier re searches and the new variables which can be accurately calculated [51]. Significant explanatory variables found in previous studies and their impacts on accident frequency are discussed in the following paragraphs.

2.3.1. ExposureVariables. The main exposure variables selected are Traffic volume and segment length [27]. Almost all researchers found traffic volume as a significant independent variable affecting accident frequency. Most of the researchers [7, 16, 17, 20, 28, 30, 40–43, 46, 47, 52–55] found that annual average daily traffic (AADT) or ADT was positively correlated with accident frequency. Zeng et al. [18] used monthly total traffic (MTT) instead of AADT to avoid information loss in time-varying variable and found positive association with accident frequency. Persaud and Dzbik [53] used hourly traffic volume for microscopic model to capture daily variation in traffic volume.

Many researchers [17, 18, 40, 41, 43, 46, 55] found the expressway segment length to be positively correlated with number of accidents. Wen et al. [56] used variable daily vehicle kilometer travel (DVKT) and found it positively associated with accident frequencies. Similarly, Wen et al. [11] used variable monthly vehicle kilometer travel (MVKT) and found positively correlated with the accident frequency.

2.3.2. Traffic Conditions. Traffic composition was used as an important independent variable by various researchers to identify the safety effects on expressway. Some researchers [20, 24, 31] found that higher percentage of trucks in traffic volume on expressway causes more number of accidents, and this could be attributed to the difference of speed between light vehicles and heavy vehicles. However, other researchers [11, 47, 57] found that high fraction of trucks and trailers in traffic volume resultin decreasing accident risk. This contradictory finding could be associated with cautious driving of light vehicle drivers close to heavy vehicles, better driving capabilities of heavy vehicle drivers as compared to light vehicle drivers, and increased homogeneity in traffic due to higher percentage of heavy vehicles in traffic volume that reduces traffic conflict [11, 47, 57]. This ambiguous conclusion needs more investigation in future. Speed limits for car and trucks were used by Hou et al. [37] as an independent variable and observed that a lower speed limit for trucks and high speed limit for cars may reduce truck accidents. Hosseinlou et al. [19] found a positive cor relation between speed and weekly accidents on expressway in Iran. However, "the safety effects of speed limits are still not fully understood and should be further investigated" [37].

2.3.3. Geometric Design Elements. Design consistency, horizontal alignment, and vertical alignment are the main component of expressway geometric design that affect the traffic safety. Montella et al. [58] found two design consistency measures namely operating speed consitency, and driving dynamics consistency as significant independent variables. Design consistency is the conformance of expressway's geometry with driver expectancy. Accident frequency on horizontal curves increases with increase in operating speed reduction and decrease in the difference between friction demand and friction supply [58].

Hou et al. [37] found that the accidents were significantly associated with curvature, deflection angle, curve direction, and length of horizontal curve. The curvature has been found positively correlated with accident occurrence [24, 37, 54]. Wen et al. [56] found that the curvature was positively associated with the property damage accidents only. Haq et al. [31] found a decreasing impact of horizontal curve radius on the truck related accidents. Studies also show that setting more gradual horizontal curves can reduce the number of accidents [46]. The deflection angle found positively correlated with the probability of accidents and left turn curves were observed to be more dangerous than right turn curves leading to different design of safety measures [37, 52]. The length of horizontal curve was found negatively associate with the likelihood of accident occurrence [37].

The expressway sections with steep grades found positively correlated with the accidents likelihood due to escalating speed variation between light and heavy vehicles [18, 37, 48]. However, Wen et al. [11] and Ma et al. [7] found negative correlation among vertical gradient and probability of accident occurrence. This is a contradictory conclusion as compared with the previous investigations [18, 37, 39, 48, 56] and need to be further investigated. Downgrade sections of expressway found more unsafe as compared to upgrade sections and this could be attributed to the failure of brakes of heavy vehicles due to frequent application to control speed [7, 24, 31, 37, 58]. Increase in vertical grade found more injury accidents on expressway, and this could be attributed to adverse impact on stopping sight distance on steep grades [27, 56]. Hou et al. [37] found that the vertical curves separations increase safety on expressway dueto improved sight distance.

2.3.4. Cross-section Elements. Many researchers [7, 18, 24, 30, 37, 41, 48] found positive correlation between number of lanes on expressway and accident probability due to escalating lane change opportunities and consequently rising conflict between the traffic.

Saeed et al. [37] found that lane width have negative correlation with the injury accidents. Ture Kibar et al. [14] concluded that truck involvement in accidents reduces with increased lane width. However, Mohammadi et al. [47] found positive correlation among the lane width and accident occurrence. Thus, this contradictory finding requires further investigation. Undersized and oversized lane factors were found statistically significant and positively related to accident frequency for urban expressway[55]. Lane width less than 3.25 m and more than 3.75 m were considered as undersized and oversized segments respectively [55].

Some researchers found a decreasing impact of adding the climbing lane on the accident frequency [24, 31, 37] along section with steep grades. This could be attributed to the reduction in interaction among light vehicles and heavy vehicles after adding climbing lane.

Hou et al. [37] found that increase in the median barrier offset can significantly lower the accidents as the larger median barrier offset reduces driver's tension and anxiety of hitting the barrier. It was found that the segments with median barriers witness decreased number of accidents [40, 42].

Median width was found significant in affecting safety for expressways [16]. A negative correlation was found between median width and number of accidents on expressway [27,42,54].

Both inside and outside shoulder widening were found associated with decreased accident count [30,44,46,51,52].

2.3.5. Category of Segment. Higher accident frequency was observed on the segments in service areas, tunnel and within interchanges as compared to the basic segments under similar conditions, probably caused by frequent braking, accelerating, and lane changing maneuvers on these segments [43].

2.3.6. Pavement Related Variables. Pavement condition factors significantly associated with accident frequency on expressway include distress ratio, rutting depth, and International Roughness Index (IRI) [37]. Friction and pavement type were also significantly affected accident frequency. Saeed et al. [43] found that the segments with excellent pavement condition (IRI $\leq$ 100) had a higher number of injury accidents due to associate higher speeds. Chen et al. [27] found that 1 percent increase in pavement serviceability index would decrease no-casualty accidents on expressway by 0.26 percent. Increasing IRI was found associated positively with the accident frequency [40]. Increased friction had a positive impact on the accident frequency [42]. Change in pavement type (from concrete to asphalt) almost always increases the probability of accidents [42]. No explanation of this phenomenon is found in the literature and hence need to be investigated further.

2.3.7. Some Other Variables. Traffic Sign: Installation of more traffic signs up to a specific number would have positive impact on accident frequency [46].

Ramp Density: Ramp density and ramp combination are statistically significant and positively related to accident frequency for urban expressway [55].

Lighting: In comparison to both side lighting, point lighting, no lighting, and continuous lighting were found to have a negative effect on accidents [41]. However, the effect of this variable was less studied.

Weather Related Variables: Haq et al. [31] found a increasing impact of rainy day per year on the truck related accidents on expressway. Strong winds increased the problem of controlling heavy vehicles on curves and consequently the frequency of accidents [37].

Pavement Marking: Zhao et al. [59] found raised pavement marking as an effective measure for reducing accidents in expressway tunnels. Lee et al. [60] observed a significant reduction in the average driving speed of the vehicles at the ramps of expressways by using transverse pavement marking and thus consequently found a reduction in accident frequency. Yang et al. [61] compared the effect of traditional marking, fishbone-shaped marking and edge-rate marking on the speed reduction of vehicles at the expressway's deceleration lane and found that fishbone-shaped marking was effective in reducing vehicles speed from 12.3 km/h to 15.2 km/h.

2.3.8. Interaction Effect Related Variables. It is a well known fact that the road accident occurs due to interaction of different risk factors. Some researchers [11, 30, 47] have tried to understand the interaction effect of risk factors on accident frequency. Mohammadi et al. [47] observed that accident frequency on an intersection in urban area decreases with increase in speed limit and at a speed limit of 55 mph, by increasing the number of lanes. Wang et al. [30] found that total number of rain related accidents reduce with increase in median width, but, single-vehicle rain related accidents increase. This might be due to reduced driving visibility and the driver's lane control ability, which results in median-related run-off-road accidents in presence of a wider median [30]. Interaction between curve and rain increased accident frequency on expressway and it might be attributed due to decline in skid resistance in rainy weather [11, 30]. Interaction between rain and vertical gradient had reduced the number of accidents and it might be due to alerted driving behavior under difficult climatic situations [11]. Wen et al. [11] also found that interaction of wind speed and vertical gradient could increase accident likelihood.

3. Conclusions

This paper attempts to review findings of more than 60 research articles related to accident prediction modeling on expressways to find out suitable methods of modeling, specific data requirements and issues related to data quality and quantity, and criteria for selection of a variable into the model. One of the major objectives of this work was to identify the areas where the results are still inconclusive. According to the literature review, it has been found that the most of the APMs have been developed for the expressways in the developed countries and only a few researches available on APMs for the expressways in the developing countries, which may be attributed to the late development of the expressways type facilities in these countries as compared to the developed one. However, the traffic composition, vehicle characteristics, driver behavior, and environmental characteristics in developing countries like India are very much different from those in developed countries. So, the APMs developed for expressways in developed countries could not be directly applied in developing countries.

The correlated random-parameter model is the most advanced approach in all the parametric modeling methods to date, as it can account for both the heterogeneous effects of explanatory factors across the road segments and the cross correlations among the random-parameter estimates, but at the same time this approach is computationally expensive.

Non parametric methods of accident prediction modeling for expressways have better predicting capabilities as compared to the parametric one, however, at the same time these models do not provide the quantification of the effect of various explanatory variables on the accident frequency.

Various independent variables (exposure, traffic condition, expressway design, cross-section, segment category, pavement condition, weather condition, and interaction effect related) have been used to develop the APMs for expressway during last three decades. Findings related to variables like percentage of heavy vehicles, vertical gradient, lane width are not conclusive requiring further investigation. However, safety effect of variables like speed limit, roadway lighting, pavement type, and fog are less studied in the literature, and should be investigated further. Literature review suggests that no study has been conducted so far on the accident frequency due to the effect of allowing two wheelers on the expressways. Also, the accidents related to fatigue or drowsiness also needs examination which is one of the leading causes of accidents on expressways in India.

References

1. WHO, "GLOBAL STATUS REPORT ON ROAD SAFETY 2018," 2018.
2. MoRTH, "Road Accidents in India [2019]," *Transp. Res. Wing, Minist. Road Transp. Highw. Gov. India, New Delhi, India.*, 2020, [Online]. Available: https://morth.nic.in/road-accident-in-india.
3. D. Lord and F. Mannering, "The statistical analysis of crash-frequency data: A review and assessment of methodological alternatives," *Transp. Res. Part A Policy Pract.*, vol. 44, no. 5, pp. 291–305, 2010, doi: 10.1016/j.tra.2010.02.001.
4. F. L. Mannering, V. Shankar, and C. R. Bhat, "Unobserved heterogeneity and the statistical analysis of highway accident data," *Anal. Methods Accid. Res.*, vol. 11, pp. 1–16, 2016, doi: 10.1016/j.amar.2016.04.001.
5. G. Singh, M. Pal, Y. Yadav, and T. Singla, "Deep neural network-based predictive modeling of road accidents," *Neural Comput. Appl.*, vol. 8, 2020, doi: 10.1007/s00521-019-04695-8.
6. G. Singh, S. N. Sachdeva, and M. Pal, "Comparison of three parametric and machine learning approaches for modeling accident severity on non-urban sections of Indian highways," *Adv. Transp. Stud.*, vol. 45, pp. 123–140, 2018.
7. Z. Ma, H. Zhang, S. I. J. Chien, J. Wang, and C. Dong, "Predicting expressway crash frequency using a random effect negative binomial model: A case study in China," *Accid. Anal. Prev.*, vol. 98, no. January, pp. 214–222, 2017, doi: 10.1016/j.aap.2016.10.012.
8. D. Mohan, G. Tiwari, and K. Bhalla, "Road safety in India: Status report 2017," *Transp. Res. Inj. Prev. Program. Indian Inst. Technol. Delhi.*, 2017, [Online]. Available: www.iitd.ac.in/-tripp.
9. F. La Torre, M. Meocci, L. Domenichini, V. Branzi, N. Tanzi, and A. Paliotto, "Development of an accident prediction model for Italian freeways," *Accid. Anal. Prev.*, vol. 124, no. March, pp. 1–11, 2019, doi: 10.1016/j.aap.2018.12.023.
10. J. Wang, Y. Kong, and T. Fu, "Expressway crash risk prediction using back propagation neural network : A brief investigation on safety resilience," *Accid. Anal. Prev.*, vol. 124, no. January, pp. 180–192, 2019, doi: 10.1016/j.aap.2019.01.007.
11. H. Wen, X. Zhang, Q. Zeng, and N. N. Sze, "Bayesian spatial-temporal model for the main and interaction effects of roadway and weather characteristics on freeway crash incidence," *Accid. Anal. Prev.*, vol. 132, no. November, p. 105249, 2019, doi: 10.1016/j.aap.2019.07.025.
12. Q. Zeng, W. Gu, X. Zhang, H. Wen, J. Lee, and W. Hao, "Analyzing freeway crash severity using a Bayesian spatial generalized ordered logit model with conditional autoregressive priors," *Accid. Anal. Prev.*, vol. 127, no. June, pp. 87–95, 2019, doi: 10.1016/j.aap.2019.02.029.
13. JP Research India Pvt Ltd, "Mumbai – Pune Expressway Road Accident Study," 2019. [Online]. Available: http://www.jpri.in/publications/.
14. P. Parida, H. S. Nagar, and N. Chakravarty, "Causes of accidents on Yamuna Expressway (Noida to Greater Noida) and suggeted remedial measures," *BVAAP*, vol. 20, no. 2, pp. 212–220, 2012, [Online]. Available: http://hdl.handle.net/123456789/18663.
15. L. S. Bisht and G. Tiwari, "Outcomes of Road Safety Audit of Yamuna Expressway," *IIT Delhi Industry Day*, 2019. .
16. M. A. Hadi, J. Aruldhas, L. F. Chow, and J. A. Wattleworth, "Estimating safety effects of cross-section design for various highway types using negative binomial regression," *Transp.*

Res. Rec. 1500, TRB, Natl. Res. Counc. Washington, D.C., pp. 169–177, 1995.

17. C. Caliendo, M. Guida, and A. Parisi, "A crash-prediction model for multilane roads," *Accid. Anal. Prev.*, vol. 39, pp. 657–670, 2007, doi: 10.1016/j.aap.2006.10.012.
18. Q. Zeng, J. Sun, and H. Wen, "Bayesian Hierarchical Modeling Monthly Crash Counts on Freeway Segments with Temporal Correlation," *J. Adv. Transp.*, vol. 2017, no. 24 October, 2017, doi: https://doi.org/10.1155/2017/5391054.
19. M. H. Hosseinlou, A. Mahdavi, and M. J. Nooghabi, "Validation of the influencing factors associated with traffic violations and crashes on freeways of developing countries : A case study of Iran," *Accid. Anal. Prev.*, vol. 121, no. December, pp. 358–366, 2018, doi: 10.1016/j.aap.2018.06.009.
20. F. Ture Kibar, F. Celik, and F. Wegman, "Analyzing truck accident data on the interurban road Ankara–Aksaray–Eregli in Turkey: Comparing the performances of negative binomial regression and the artificial neural networks models," *J. Transp. Saf. Secur.*, vol. 11, no. 2, pp. 129–149, 2019, doi: 10.1080/19439962.2017.1363841.
21. D. Kim and Y. Lee, "Modelling crash frequencies at signalized intersections with a truncated count data model," *Int. J. Urban Sci.*, vol. 17, no. 1, pp. 85–94, 2013, doi: 10.1080/12265934.2013.774702.
22. Y. Zou, L. Wu, and D. Lord, "Modeling over-dispersed crash data with a long tail : Examining the accuracy of the dispersion parameter in Negative Binomial models," *Anal. Methods Accid. Res.*, vol. 5–6, pp. 1–16, 2015, doi: 10.1016/j.amar.2014.12.002.
23. M. Shirazi, D. Lord, S. S. Dhavala, and S. R. Geedipally, "A semiparametric negative binomial generalized linear model for modeling over-dispersed count data with a heavy tail : Characteristics and applications to crash data," *Accid. Anal. Prev.*, vol. 91, pp. 10–18, 2016, doi: 10.1016/j.aap.2016.02.020.
24. Q. Hou, X. Meng, J. Leng, and L. Yu, "Application of a random effects negative binomial model to examine crash frequency for freeways in China," *Physica A*, vol. 509, pp. 937–944, 2018, doi: 10.1016/j.physa.2018.06.088.
25. K. Wang, J. N. Ivan, N. Ravishanker, and E. Jackson, "Multivariate poisson lognormal modeling of crashes by type and severity on rural two lane highways," *Accid. Anal. Prev.*, vol. 99, pp. 6–19, 2017, doi: 10.1016/j.aap.2016.11.006.
26. Z. Li, W. Wang, P. Liu, L. Bai, and M. Du, "Analysis of Crash Risks by Collision Type at Freeway Diverge Area Using Multivariate Modeling Technique," *J. Transp. Eng.*, vol. 141, no. 6, 2015, doi: 10.1061/(ASCE)TE.1943-5436.0000757.
27. S. Chen, T. Usman, M. Alinizzi, S. Lavrenz, and S. Labi, "Safety sensitivity to roadway characteristics : A comparison across highway classes," *Accid. Anal. Prev.*, vol. 123, pp. 39–50, 2019, doi: 10.1016/j.aap.2018.10.020.
28. P. C. Anastasopoulos, "Random parameters multivariate tobit and zero-inflated count data models : Addressing unobserved and zero-state heterogeneity in accident injury-severity rate and frequency analysis," *Anal. Methods Accid. Res.*, vol. 11, pp. 17–32, 2016, doi: 10.1016/j.amar.2016.06.001.
29. E. S. Park and D. Lord, "Multivariate Poisson – Lognormal Models for Jointly Modeling Crash Frequency by Severity," *Transp. Res. Rec. J. Transp. Res. Board, No. 2019, Transp. Res. Board Natl. Acad. Washington, D.C.*, pp. 1–6, 2007, doi: 10.3141/2019-01.
30. K. Wang, S. Zhao, and E. Jackson, "Multivariate Poisson Lognormal Modeling of Weather-

Related Crashes on Freeways," *Transp. Res. Rec. J. Transp. Res. Board, No. 2672, Transp. Res. Board Natl. Acad. Washington, D.C.*, no. 38, pp. 184–198, 2018, doi: 10.1177/0361198118776523.

31. M. T. Haq, M. Zlatkovic, and K. Ksaibati, "Freeway Truck Traffic Safety in Wyoming : Crash Characteristics and Prediction Models," *Transp. Res. Rec. J. Transp. Res. Board, No. 2673, Transp. Res. Board Natl. Acad. Washington, D.C.*, no. 10, pp. 333–342, 2019, doi: 10.1177/0361198119847980.
32. D. Lord, S. P. Washington, and J. N. Ivan, "Poisson , Poisson-gamma and zero-inflated regression models of motor vehicle crashes : balancing statistical fit and theory," *Accid. Anal. Prev.*, vol. 37, pp. 35–46, 2005, doi: 10.1016/j.aap.2004.02.004.
33. C. Liu, M. Zhao, W. Li, and A. Sharma, "Multivariate random parameters zero-inflated negative binomial regression for analyzing urban midblock crashes," *Anal. Methods Accid. Res.*, vol. 17, pp. 32–46, 2018, doi: 10.1016/j.amar.2018.03.001.
34. G. Singh, S. N. Sachdeva, and M. Pal, "M5 model tree based predictive modeling of road accidents on non-urban sections of highways in India," *Accid. Anal. Prev.*, vol. 96, pp. 108–117, 2016, doi: 10.1016/j.aap.2016.08.004.
35. M. Quddus, "Exploring the Relationship Between Average Speed , Speed Variation , and Accident Rates Using Spatial Statistical Models and GIS Exploring the Relationship Between Average Speed , Speed Variation , and Accident Rates Using Spatial Statistical Models and G," *J. Transp. Saf. Secur.*, vol. 5, no. 1, pp. 27–45, 2013, doi: 10.1080/19439962.2012.705232.
36. G. I. M. A. Mothafer, T. Yamamoto, and V. N. Shankar, "A negative binomial crash sum model for time invariant heterogeneity in panel crash data : Some insights," *Anal. Methods Accid. Res.*, vol. 14, no. June, pp. 1–9, 2017, doi: 10.1016/j.amar.2016.12.003.
37. Q. Hou, A. P. Tarko, and X. Meng, "Investigating factors of crash frequency with random effects and random parameters models : New insights from Chinese freeway study," *Accid. Anal. Prev.*, vol. 120, no. November, pp. 1–12, 2018, doi: 10.1016/j.aap.2018.07.010.
38. Q. Hou, A. P. Tarko, and X. Meng, "Analyzing crash frequency in freeway tunnels : A correlated random parameters approach," *Accid. Anal. Prev.*, vol. 111, pp. 94–100, 2018, doi: 10.1016/j.aap.2017.11.018.
39. M. Ahmed, H. Huang, M. Abdel-aty, B. Guevara, and M. Marker, "Exploring a Bayesian hierarchical approach for developing safety performance functions for a mountainous freeway," *Accid. Anal. Prev.*, vol. 43, pp. 1581–1589, 2011, doi: 10.1016/j.aap.2011.03.021.
40. P. C. Anastasopoulos and F. L. Mannering, "A note on modeling vehicle accident frequencies with random-parameters count models," *Accid. Anal. Prev.*, vol. 41, pp. 153–159, 2009, doi: 10.1016/j.aap.2008.10.005.
41. N. S. Venkataraman, G. F. Ulfarsson, V. Shankar, J. Oh, and M. Park, "Model of Relationship Between Interstate Crash Occurrence and Geometrics Negative Binomial Approach," *Transp. Res. Rec. J. Transp. Res. Board, No. 2236, Transp. Res. Board Natl. Acad. Washington, D.C.*, pp. 41–48, 2011, doi: 10.3141/2236-05.
42. M. R. Rahman Shaon, X. Qin, M. Shirazi, D. Lord, and S. R. Geedipally, "Developing a Random Parameters Negative Binomial-Lindley Model to analyze highly over-dispersed crash count data," *Anal. Methods Accid. Res.*, vol. 18, no. June, pp. 33–44, 2018, doi: 10.1016/j.amar.2018.04.002.

43. T. U. Saeed, T. Hall, H. Baroud, and M. J. Volovski, "Analyzing road crash frequencies with uncorrelated and correlated random-parameters count models : An empirical assessment of multilane highways," *Anal. Methods Accid. Res.*, vol. 23, no. September, 2019, doi: 10.1016/j.amar.2019.100101.
44. J. Liu, A. J. Khattak, and B. Wali, "Do safety performance functions used for predicting crash frequency vary across space? Applying geographically weighted regressions to account for spatial heterogeneity," *Accid. Anal. Prev.*, vol. 109, no. December, pp. 132–142, 2017, doi: 10.1016/j.aap.2017.10.012.
45. AASHTO, "Highway Safety Manual, first edition. 2014 Supplement," *Am. Assoc. State Highw. Transp. Off.*, 2014.
46. L. Zheng, J. Sun, and X. Meng, "Crash prediction model for basic freeway segments incorporating influence of road geometrics and traffic signs," *J. Transp. Eng. Part A Syst.*, vol. 144, no. 7, 2018, doi: 10.1061/JTEPBS.0000155.
47. M. A. Mohammadi, V. A. Samaranayake, and G. H. Bham, "Crash frequency modeling using negative binomial models : An application of generalized estimating equation to longitudinal data," *Anal. Methods Accid. Res.*, vol. 2, pp. 52–69, 2014, doi: 10.1016/j.amar.2014.07.001.
48. L. Chang, "Analysis of freeway accident frequencies : Negative binomial regression versus artificial neural network," *Saf. Sci.*, vol. 43, pp. 541–557, 2005, doi: 10.1016/j.ssci.2005.04.004.
49. Y. Xie, D. Lord, and Y. Zhang, "Predicting motor vehicle collisions using Bayesian neural network models : An empirical analysis," *Accid. Anal. Prev.*, vol. 39, pp. 922–933, 2007, doi: 10.1016/j.aap.2006.12.014.
50. L. Chang and W. Chen, "Data mining of tree-based models to analyze freeway accident frequency," *J. Safety Res.*, vol. 36, pp. 365–375, 2005, doi: 10.1016/j.jsr.2005.06.013.
51. R. Eenink, M. Reurings, R. Elvik, J. Cardoso, S. Wichert, and C. Stefan, "Accident Prediction Models and Road Safety Impact Assessment: recommendations for using these tools. Report RI-SWOV-WP2-D2-F. SWOV Institute for Road Safety Research, Leidschendam," 2007.
52. Q. Hou, X. Huo, and J. Leng, "A correlated random parameters tobit model to analyze the safety effects and temporal instability of factors affecting crash rates," *Accid. Anal. Prev.*, vol. 134, no. 105326, 2020, doi: 10.1016/j.aap.2019.105326.
53. B. Persaud and L. Dzbik, "Accident prediction models for freeways," *Transp. Res. Rec. 1401, TRB, Natl. Res. Counc. Washington, D.C.*, vol. 1401, pp. 55–60, 1993.
54. B.-J. Park, K. Fitzpatrick, and D. Lord, "Evaluating the effects of freeway design elements on safety," *Transp. Res. Rec. 2195, TRB, Natl. Res. Counc. Washington, D.C.*, no. 1, pp. 58–69, 2010, doi: 10.3141/2195-07.
55. L. Wu, J. Sun, and T. Li, "Relationship between Lane Width and Safety along Urban Expressways in Shanghai," *J. Transp. Eng. Part A Syst.*, vol. 145, no. 3, 2019, doi: 10.1061/ JTEPBS.0000216.
56. H. Wen, J. Sun, Q. Zeng, X. Zhang, and Q. Yuan, "The effects of traffic composition on freeway crash frequency by injury severity: A Bayesian multivariate spatial modeling approach," *J. Adv. Transp.*, 2018, doi: https://doi.org/10.1155/2018/6964828.
57. Q. Zeng and H. Huang, "Bayesian spatial joint modeling of traffic crashes on an urban road network," *Accid. Anal. Prev.*, vol. 67, pp. 105–112, 2014, doi: 10.1016/j.aap.2014.02.018.
58. A. Montella, L. Colantuoni, and R. Lamberti, "Crash prediction models for rural motorways,"

Transp. Res. Rec. J. Transp. Res. Board, No. 2083, Transp. Res. Board Natl. Acad. Washington, D.C., 2008, pp. 180–189, 2008, doi: 10.3141/2083-21.

59. X. Zhao, Y. Ju, H. Li, C. Zhang, and J. Ma, "Safety of Raised Pavement Markers in Freeway Tunnels Based on Driving Behavior," *Accid. Anal. Prev.*, vol. 145, no. July, 2020, doi: 10.1016/j.aap.2020.105708.
60. S. M. Lee, K. Lee, and S. Kim, "Effectiveness of the Peripheral Transverse Line as Speed-Reduction Treatment on Korean Expressway Ramps," *Int. J. Highw. Eng.*, vol. 14, no. 6, pp. 85–92, 2012, [Online]. Available: http://www.ksre.or.kr/.
61. Y. Yang, S. M. Easa, X. Zheng, A. Hu, F. Liu, and M. Chen, "Evaluation effects of two types of freeway deceleration markings in China," *PLoS One*, vol. 14, no. 8, pp. 1–18, 2019, doi: 10.1371/journal.pone.0220811.

AUTHOR INDEX